# Advanced Sensing Technology for Ocean Observation

# Advanced Sensing Technology for Ocean Observation

Guest Editor

**Federico Angelini**

MDPI

Basel • Beijing • Wuhan • Barcelona • Belgrade • Novi Sad • Cluj • Manchester

*Guest Editor*
Federico Angelini
Laboratory FSN-TECFIS-DIM
ENEA
Frascati
Italy

*Editorial Office*
MDPI AG
Grosspeteranlage 5
4052 Basel, Switzerland

This is a reprint of the Special Issue, published open access by the journal *Sensors* (ISSN 1424-8220), freely accessible at: https://www.mdpi.com/journal/sensors/special_issues/ocean_sensing.

For citation purposes, cite each article independently as indicated on the article page online and as indicated below:

Lastname, A.A.; Lastname, B.B. Article Title. *Journal Name* **Year**, *Volume Number*, Page Range.

**ISBN 978-3-7258-4355-8 (Hbk)**
**ISBN 978-3-7258-4356-5 (PDF)**
**https://doi.org/10.3390/books978-3-7258-4356-5**

Cover image courtesy of Federico Angelini

# Contents

# About the Editor

**Federico Angelini**

Born in Rome in 1975, Angelini earned a degree in Physics from the University of Rome "La Sapienza" in 2001, specializing in terrestrial physics with a thesis on Raman spectroscopy for atmospheric water vapor detection. He later completed a PhD in remote sensing (2005), focusing on atmospheric aerosol and humidity measurements. Since then, he has been actively engaged in Earth remote sensing, specializing in atmospheric and ocean composition analysis, primarily using inelastic lidar techniques such as Raman and fluorescence. His expertise covers trace detection applications in cultural heritage sciences, security, and forensics. Since 2012, he has worked as a researcher at the Italian National Agency for New Technologies, Energy, and Sustainable Development (ENEA), contributing to advancements in diagnostics and metrology.

# Preface

The oceans are crucial for both climate change and sustainable development, as they interact with the atmosphere through complex feedback, and tipping points. Human activities such as pollution, transport, and fisheries further impact ocean health. Scientists and policymakers agree that a deeper understanding of marine ecosystems is essential for effective mitigation and adaptation strategies. However, many aspects of ocean science remain unexplored due to the complexity of physical, chemical, and biological interactions and the challenges of obtaining continuous, accurate measurements. While satellite-based remote sensing offers wide coverage, it has limitations, particularly for subsurface profiling. Addressing oceanographic challenges requires reliable data across various spatial and temporal scales, from molecular processes to global climate dynamics. This Special Issue compiles innovative research on signal processing and machine learning techniques to enhance ocean data quality, resolution, and integration, ultimately advancing our understanding of ocean–climate interactions and informing better strategies for sustainability.

**Federico Angelini**
*Guest Editor*

sensors

MDPI

*Editorial*

# Advanced Sensing Technology for Ocean Observation

**Federico Angelini**

Italian National Agency for New Technologies, Energy and Sustainable Economic Development (ENEA), Laboratory FSN-TECFIS-DIM, 00044 Frascati, Italy; federico.angelini@enea.it

## 1. Introduction

It is almost impossible to overestimate the importance of the oceans for human society and the whole biosphere, either from the perspectives of climate change or sustainable development [1]. On the one hand, reciprocal interactions between the atmosphere and the oceans are recognized to have a key role in the climate change and to act through very complex feedbacks [2,3] and tipping points [4]. On the other hand, sustainable development is tightly connected to a conscious exploitation of the oceans [5–7]; just think of the impact of pollution, transport, and fisheries on human health and the whole society.

It is widely agreed among the scientific community, as well as among policymakers, that a deep understanding of the ocean ecosystem is required to create figure out and implement mitigation and adaptation plans [8].

It has long been known that the oceans play a key role in determining climate conditions and life on the Earth [9]. For many decades, scientists have been raising the alarm on the health of the oceans [10] and their impact on climate change [11]. Marine ecosystems are changing quickly, at unprecedented rates [12] in terms of physical, chemical, and biological characteristics [13]. Nevertheless, it is not easy to detect and quantify all these changes [14].

In fact, despite its importance, many aspects of sea science are still relatively unknown [15], both because of the complexity of interactions between physics, chemistry, biology, and human activities, and because of the difficulties inherent in carrying out accurate, continuous, and reliable measurements of even simple variables such as temperature and salinity. Many kinds of remote and/or automated measurements, conducted using technology such as satellites, offer great advantages in terms of coverage and continuity, but suffer important drawbacks that cannot be overcome through other satellite observations (for example, vertical profiles in water are inaccessible using remote sensing because of the strong absorption of electromagnetic waves by water).

Many current problems, spanning different spatial scales (from molecular processes to global issues) and timescales (from chemical dynamics to geological eras), still limit our understanding of the processes that involve ocean physics, chemistry, and biology. Accurate, reliable, and continuous measurements in a variety of fields [16–18], from basic physics to applications for the optimization and sustainability of fishing and navigation, present new challenges. This Special Issue aimed to gather valuable and innovative papers on a wide range of new methods and technologies to improve the quality and resolution of oceanographic data, and to integrate different data sources.

## 2. Overview of Published Papers

This Special Issue gathers together nine research papers and one technical note, discussing new progresses in marine sensing in a wide range of technologies. These are summarized below, sorted by publication date.

In Contribution 1, Zhang et al. present a new method for creating mosaics from striped images obtained by side-scan sonar technology. The problem addressed is the distortion and different resolution of overlapping areas in images, typical of complex marine environments. The proposed solution is based on curvelet transform, a multi-scale analysis method more effective than the wavelet transform for the representation of details and edges in images. First, a register of the images is made to eliminate distortions and dislocations, and then a resolution vector for overlapping areas is calculated and a resolution weight model is used to guide the fusion of the curvelet coefficients. Eventually, the inverse curvelet transform generates the final mosaic. Experiments with real data demonstrate the superiority of the proposed method over traditional image fusion algorithms.

In Contribution 2, Jutard et al. describe a new quality control protocol ("delayed-mode quality control") for radiometry data acquired by the Biogeochemical-Argo (BGC-Argo) floats. The main problem addressed is the correction of systematic errors in data due to the temperature dependence and time drift of irradiance sensors. The authors propose a method to correct these errors using auxiliary night measurements and daily measurements at 1000 dbar, validating the method on over 10,000 profiles from various ocean regions. The aim is to improve the quality of BGC-Argo's radiometry data, making them more accurate and reliable for oceanographic research and, in particular, for studying phytoplankton dynamics and integrating with satellite observations.

In Contribution 3, Wassim Baba et al. present, as a technical note, an innovative methodology for obtaining large-scale coastal bathymetry using Sentinel-2 satellite imagery and a high-performance cluster (HPC). The main objective is to overcome the limitations of traditional, time-consuming, and costly techniques for mapping coastal waters. The researchers describe an approach that uses the properties of ocean waves, visible in Sentinel-2 images, to estimate water depth, implementing it on an HPC to handle the high volume of data. The case study focuses on the North African coast, comparing results obtained with a reference bathymetric product (GEBCO), highlighting the potential of the method to create high-resolution global bathymetric maps. The methodology therefore offers a more efficient and cost-effective approach to monitoring coastal areas, which are crucial for the environment and economic development.

In Contribution 4, Li et al. describe an innovative method for estimating water transport in an Arctic lagoon connected to the Beaufort Sea. Using a combination of short-term measurements from a ship-mounted acoustic Doppler current profiler (ADCP) and long-term measurements from a bottom-anchored ADCP, the researchers established a statistical relationship between the measured water velocity and total transport. This approach, validated with a coefficient of determination R2 of 0.89, allows the estimation of water transport over longer periods, overcoming the constraints imposed by the harsh environmental conditions of the Arctic and limited resources. The study highlights the importance of measuring water transport in the Arctic regions to understand the impact of climate change and provides an effective and cost-effective methodology for addressing this challenge.

In Contribution 5, Wang et al. present an evaluation of a new lightweight mouse-bathymetric LiDAR system mounted on a drone (UAV), called Mapper4000U. The study compares the bathymetric performance of the Mapper4000U with that of a LiDAR system mounted on a manned aircraft, using data from a Chinese coastal island and data from a multibeam sonar as a reference. The main objective is to demonstrate the Mapper4000U's ability to perform high-resolution bathymetric mapping in shallow water, including underwater object detection. The results show a high precision and accuracy of the UAV system, with a significantly higher point density than the traditional LiDAR system, while maintaining a good penetration depth. The article examines in detail the data processing methodology and discusses the environmental effects on measurements.

In Contribution 6, Nekrasov et al. present research on the optimization of sea wind measurements using an aerial scatterometer with a rotating antenna mounted under the fuselage. The study focuses on the analysis of normalized radar cross-section sampling at different incidence angles, assessing the accuracy of wind vector estimation using a single angle or combinations of angles. The results of the Monte Carlo simulations show that using more than one close angle of incidence significantly improves measurement accuracy while reducing maximum operating altitude. The work aims to improve the functionality of existing airborne radar and to drive the development of new remote sensing systems for measuring sea wind.

In Contribution 7, Nagano et al. describe an experiment conducted off the coast of Sanriku, Japan to study the turbulent heat flux between the ocean and the atmosphere. Using an autonomous marine vehicle, a Wave Glider, the researchers measured various parameters (air and sea temperature, humidity, wind speed) for 55 days to calculate heat flow, focusing in particular on sub-mesoscale variations. The study shows that the intrusion of cold and dry air masses, following the passage of low atmospheric pressure systems, generates a significant heat flux upwards over relatively warmer water regions, highlighting the importance of high-resolution observations to fully understand the ocean–atmosphere interaction and improve weather and climate forecasts. The results show that satellite observations, due to their low resolution, may underestimate the influence of sub-mesoscale variations in heat flux.

In Contribution 8, Hoffman et al. present a new, simple, and economical method for estimating the coefficient of resistance of submerged floats, in particular focusing on those with complex and non-rigid shapes, which are difficult to model using analytical or numerical methods. The authors propose an in-situ approach based on the relationship between the float's attitude speed and its weight, varied by adding ballast. The accuracy of the method in obtaining the coefficient of resistance, hydrostatic strength, and, if present, the force of the thruster is demonstrated by experiments at sea. The simplicity of the method makes it applicable to arbitrarily shaped objects, overcoming the limitations and costs of conventional methods such as CFD simulations or dry-dock tests.

In Contribution 9, Nie et al. describe the development and use of a long-range hybrid Autonomous Underwater Vehicle (AUV) to measure ocean turbulence. The unique AUV combines the features of a traditional AUV and a variable-floating glider, offering different flexible movement modes. It was deployed for continuous measurements in the northern part of the South China Sea, collecting high spatial and temporal resolution data on turbulence and its relationship with thermocline, highlighting how the latter acts as a "barrier" against the transmission of energy from the surface level to greater depths. The study shows the usefulness of these hybrid AUVs as a powerful tool for studying ocean turbulence on a large scale.

In Contribution 10, Sattar present a new autonomous acoustic method for the identification of vocalizations of endangered whales, focusing on blue whales and common whales. The proposed method combines wavelet scattering transform with a deep-learning LSTM classifier, demonstrating high classification accuracy (over 97%) even with limited data sets. This represents a significant improvement over existing methods, making it possible to monitor the acoustic performance of these species more efficiently for conservation purposes, allowing better tracking of their numbers, migratory routes, and habitats. The research highlights the importance of artificial intelligence and deep learning in marine acoustic data analysis for whale protection.

## 3. Conclusions

The published papers cover a range of applications of signal processing and machine learning techniques for the analysis of data from marine environments. The works also highlight the importance of developing new methods and technologies to improve the quality and resolution of oceanographic data, as well as integrating different data sources for a more complete understanding of ocean processes.

The wide range of topics discussed in this Special Issue bears witness to the importance of improvements in ocean monitoring and the need for a better understanding of all the processes involving biology, chemistry, and physics in the oceans to better understand, in turn, connections with climate change and to plan the best strategies for mitigation and adaptation.

**Funding:** This research received no external funding.

**Conflicts of Interest:** The author declares no conflict of interest.

**List of Contributions:**

1. Zhang, N.; Jin, S.; Bian, G.; Cui, Y.; Chi, L. A Mosaic Method for Side-Scan Sonar Strip Images Based on Curvelet Transform and Resolution Constraints. *Sensors* **2021**, *21*, 6044. https://doi.org/10.3390/s21186044.
2. Jutard, Q.; Organelli, E.; Briggs, N.; Xing, X.; Schmechtig, C.; Boss, E.; Poteau, A.; Leymarie, E.; Cornec, M.; D'Ortenzio, F.; et al. Correction of Biogeochemical-Argo Radiometry for Sensor Temperature-Dependence and Drift: Protocols for a Delayed-Mode Quality Control. *Sensors* **2021**, *21*, 6217. https://doi.org/10.3390/s21186217.
3. Baba, M.W.; Thoumyre, G.; Bergsma, E.W.J.; Daly, C.J.; Almar, R. Deriving Large-Scale Coastal Bathymetry from Sentinel-2 Images Using an HIGH-Performance Cluster: A Case Study Covering North Africa's Coastal Zone. *Sensors* **2021**, *21*, 7006. https://doi.org/10.3390/s21217006.
4. Li, C.; Boswell, K.M. Estimating Water Transport from Short-Term Vessel-Based and Long-Term Bottom-Mounted Acoustic Doppler Current Profiler Measurements in an Arctic Lagoon Connected to the Beaufort Sea. *Sensors* **2022**, *22*, 68. https://doi.org/10.3390/s22010068.
5. Wang, D.; Xing, S.; He, Y.; Yu, J.; Xu, Q.; Li, P. Evaluation of a New Lightweight UAV-Borne Topo-Bathymetric LiDAR for Shallow Water Bathymetry and Object Detection. *Sensors* **2022**, *22*, 1379. https://doi.org/10.3390/s22041379.
6. Nekrasov, A.; Khachaturian, A.; Vorobev, E. Optimization of the NRCS Sampling at the Sea Wind Retrieval by the Airborne Rotating-Beam Scatterometer Mounted under Fuselage. *Sensors* **2022**, *22*, 4016. https://doi.org/10.3390/s22114016.
7. Nagano, A.; Hasegawa, T.; Ariyoshi, K.; Iinuma, T.; Fukuda, T.; Fujii, N.; Tomita, F.; Hino, R. USV-Observed Turbulent Heat Flux Induced by Late Spring Cold Dry Air Incursion over Sub-Mesoscale Warm Regions off Sanriku, Japan. *Sensors* **2022**, *22*, 9695. https://doi.org/10.3390/s22249695.
8. Hoffman, Y.; Nagar, L.; Shachar, I.; Diamant, R. A Simple Approach to Estimate the Drag Coefficients of a Submerged Floater. *Sensors* **2023**, *23*, 1394. https://doi.org/10.3390/s23031394.
9. Nie, Y.; Luan, X.; Huang, Y.; Du, L.; Song, D.; Liu, X. Microstructure Turbulence Measurement in the Northern South China Sea from a Long-Range Hybrid AUV. *Sensors* **2023**, *23*, 2014. https://doi.org/10.3390/s23042014.
10. Sattar, F. A New Acoustical Autonomous Method for Identifying Endangered Whale Calls: A Case Study of Blue Whale and Fin Whale. *Sensors* **2023**, *23*, 3048. https://doi.org/10.3390/s23063048.

## References

1. Duarte, C.M. Global change and the future ocean: A grand challenge for marine sciences. *Front. Mar. Sci.* **2014**, *1*, 63. [CrossRef]
2. Zhang, Y.; Perrie, W. Feedback mechanisms for the atmosphere and ocean surface. *Bound. Layer Meteorol.* **2001**, *100*, 321–348. [CrossRef]

3. Soden, B.J.; Held, I.M. An assessment of climate feedbacks in coupled ocean–atmosphere models. *J. Clim.* **2006**, *19*, 3354–3360. [CrossRef]
4. Heinze, C.; Blenckner, T.; Martins, H.; Rusiecka, D.; Döscher, R.; Gehlen, M.; Gruber, N.; Holland, E.; Hov, Ø.; Joos, F.; et al. The quiet crossing of ocean tipping points. *Proc. Natl. Acad. Sci. USA* **2021**, *118*, e2008478118. [CrossRef] [PubMed]
5. Mann-Borgese, E. Sustainable development in the oceans. *Environ. Policy Law* **1997**, *27*, 203.
6. Hanson, A.J. Sustainable development and the oceans. *Ocean Coast. Manag.* **1998**, *39*, 167–177. [CrossRef]
7. Pandey, U.C.; Nayak, S.R.; Roka, K.; Jain, T.K. Oceans and sustainable development. In *SDG14–Life Below Water: Towards Sustainable Management of Our Oceans*; Emerald Publishing Limited: Bingley, UK, 2021; pp. 17–42.
8. Wood, R.A.; Baker, J.A.; Beaugrand, G.; Boutin, J.; Conversi, A.; Donner, R.V.; Frenger, I.; Goberville, E.; Hayashida, H.; Koeve, W.; et al. Opportunities for Earth observation to inform risk management for ocean tipping points. *Surv. Geophys.* **2024**, 1–60. [CrossRef]
9. Littlehales, G.W. The ocean among the factors of the control of climate, Eos. *Trans. Am. Geophys. Union* **1927**, *8*, 26–31.
10. Raper, S.C.B.; Wigley, T.M.L.; Jones, P.D.; Kelly, P.M.; Mayes, P.R.; Limbert, D.W.S. Recent temperature changes in the Arctic and Antarctic. *Nature* **1983**, *306*, 458–459. [CrossRef]
11. Reid, P.C.; Fischer, A.C.; Lewis-Brown, E.; Meredith, M.P.; Sparrow, M.; Andersson, A.J.; Antia, A.; Bates, N.R.; Bathmann, U.; Beaugrand, G.; et al. Impacts of the oceans on climate change. *Adv. Mar. Biol.* **2009**, *56*, 1–150.
12. Halpern, B.S.; Frazier, M.; Afflerbach, J.; Lowndes, J.S.; Micheli, F.; O'hara, C.; Scarborough, C.; Selkoe, K.A. Recent pace of change in human impact on the world's ocean. *Sci. Rep.* **2019**, *9*, 11609. [CrossRef] [PubMed]
13. Jewett, L.; Romanou, A. Ocean acidification and other ocean changes. *Clim. Sci. Spec. Rep. Fourth Natl. Clim. Assess.* **2017**, *1*, 364–392.
14. Karl, T.R.; Derr, V.E.; Easterling, D.R.; Folland, C.K.; Hofmann, D.J.; Levitus, S.; Nicholls, N.; Parker, D.E.; Withee, G.W. Critical issues for long-term climate monitoring. *Clim. Change* **1995**, *31*, 185–221. [CrossRef]
15. Favali, P.; Dañobeitia, J.J.; Howe, B.M.; Ruhl, H.A. The discovery of the unknown planet: The ocean. *Front. Mar. Sci.* **2023**, *10*, 1221674. [CrossRef]
16. Weller, R.A.; Baker, D.J.; Glackin, M.M.; Roberts, S.J.; Schmitt, R.W.; Twigg, E.S.; Vimont, D.J. The challenge of sustaining ocean observations. *Front. Mar. Sci.* **2019**, *6*, 105. [CrossRef]
17. National Academies of Sciences; Division on Earth, Life Studies; Board on Atmospheric Sciences, Ocean Studies Board; Committee on Sustaining Ocean Observations to Understand Future Changes in Earth's Climate. *Sustaining Ocean Observations To Understand Future Changes in Earth's Climate*; National Academies Press: Cambridge, MA, USA, 2018.
18. Levin, L.A.; Bett, B.J.; Gates, A.R.; Heimbach, P.; Howe, B.M.; Janssen, F.; McCurdy, A.; Ruhl, H.A.; Snelgrove, P.; Stocks, K.I.; et al. Global observing needs in the deep ocean. *Front. Mar. Sci.* **2019**, *6*, 241. [CrossRef]

sensors

Article

# A New Acoustical Autonomous Method for Identifying Endangered Whale Calls: A Case Study of Blue Whale and Fin Whale

**Farook Sattar**

Department of Electrical and Computer Engineering, University of Victoria, Victoria, BC V8W 2Y2, Canada; fsattar@ieee.org

**Abstract:** In this paper, we study to improve acoustical methods to identify endangered whale calls with emphasis on the blue whale (*Balaenoptera musculus*) and fin whale (*Balaenoptera physalus*). A promising method using wavelet scattering transform and deep learning is proposed here to detect/classify the whale calls quite precisely in the increasingly noisy ocean with a small dataset. The performances shown in terms of classification accuracy (>97%) demonstrate the efficiency of the proposed method which outperforms the relevant state-of-the-art methods. In this way, passive acoustic technology can be enhanced to monitor endangered whale calls. Efficient tracking of their numbers, migration paths and habitat become vital to whale conservation by lowering the number of preventable injuries and deaths while making progress in their recovery.

**Keywords:** whale calls; marine bioacoustics; endangered whale; deep learning; artificial intelligence; wavelet scattering transform; identification; small data set

## 1. Introduction

It is of utmost importance to conserve endangered whale populations which are declining due to various reasons, such as striking by ships/vessels, entangling with fishing gear, and global warming. Automated acoustic monitoring has been used for monitoring marine species such as whales. Recorded acoustic samples let us listen and analyze sounds from soniferous whales for their identification. As sound travels quicker than light underwater, it is good to do acoustic monitoring than video surveillance for identifying endangered whales. However, accurate acoustic identification of endangered whale calls (vocalizations) is still difficult, especially when a whale population is getting dangerously small and the size of the available data samples is also too small.

Blue whales (*Balaenoptera musculus*) are the largest of the baleen whales and are endangered worldwide [1]. Blue whale calls are low-frequency (20–100 Hz) and repetitive [2]. Blue whales are known to produce downswept FM (frequency-modulated) calls that are often referred as D-calls. Both male and female blue whales have been found to produce such calls [3]. On the contrary, it is observed that only males produce song, and with that, these calls are associated with the breeding season. Thus, blue whale song apparently carries information about the population. Similarly, fin whales (*Balaenoptera physalus*) are listed as endangered species, which also produce low-frequency vocalizations (i.e., <100 Hz) [4,5]. Single vocalizations, in particular, are generated by male fin whales, whereas songs in the form of pulse trains can occur at high sound pressure that can be detected over a long distance (e.g., >50 km) [6]. In locations of high fin whale density, the songs and single vocalizations of numerous fin whales do overlap in time and frequency, producing the so-called fin whale chorus [6].

The development of robust deep learning methods to identify whales or finding when and where each whale population occurs is getting much attention. Recent abundance estimates using acoustic whale calls can aid assessment of the current status of each identified whale population, especially for blue whales and fin whales whose population sizes are critically decaying. This status assessment thus provides us the basis to a proper management plan for the conservation of endangered whale populations.

Deep learning motivation is greatly deduced by artificial intelligence (AI), which simulates the ability of the human brain in terms of analyzing, making decisions, and learning. The AI-enabled technique, such as deep learning, is quickly becoming a mainstream technology that can analyze large volumes of data and potentially unlock insights, especially in ocean monitoring applications. Deep learning can be defined as a technique of machine learning to learn useful features directly from given sounds, images, and texts. The core of deep learning is hieratically computing features and representing information, such as defining the features starting from a low level to high level. Different hidden layers are involved in making decisions by using the feedback from one layer to the previous one, or the resulting layer will have been fed into the first layer. Therefore, many layers are exploited by deep learning for nonlinear data processing of supervised or unsupervised feature extraction for classification and pattern recognition. It is difficult for a computer to understand complex data, such as an image or a sequence of data of a complex nature, so deep learning algorithms are used instead of usual learning methods. The conventional methods have been overtaken by deep learning methods which can detect and classify objects in complex scenarios. Deep learning can thus help us to create better ocean acoustic detection and classification models.

Recently, a few studies have been reported about the deep learning methods developed for blue whale and fin whale monitoring. In the Ref. [2], Siamese neural networks (SNN) were utilized to detect/classify blue whale calls from the acoustic recordings. The method classified calls from four populations of blue whales providing the highest accuracy of 94.30% for Antarctic blue whale calls, while the lowest accuracy of 90.80% was provided for the central Indian Ocean (CIO) blue calls. Studies in the Ref. [3] showed that the DenseNet-automated blue whale D-Call detector, which is based on conventional Convolutional Neural Networks (CNN) provided better results in terms of detection probability than that of human observers' analyses, particularly at low and medium signal-to-noise ratios (SNRs). Higher detection probabilities (0.905 and 0.910 for low and medium SNRs) were obtained compared to the detection probabilities obtained by human observers' analyses (0.699 and 0.697 for low and medium SNRs). In the Ref. [3], a long-term recording dataset, particularly the Australian Antarctic Division's "Casey2019" dataset was used for the results. In the Ref. [7], a two-stage deep-learning approach was developed based on a region-based convolutional neural network (rCNN) and following CNN to automatically detect/classify both blue whale D-calls and fin whale 40-Hz calls. In stage 1, the detection of regions of interests (ROIs) containing potential calls was performed using rCNN. In stage 2, a CNN was employed to classify the target whale calls from the detected ROIs. The work in the Ref. [8] presents an application of deep learning for automatic identification of fin whale 20 Hz calls, which are sometimes contaminated with other sources, such as shipping and earthquakes. All of these recently proposed advanced deep-learning-based methods have one common feature—they all require a large dataset for learning.

In this paper, taking advantage of deep nonlinear features of wavelet scattering transform (WST) [9], we adopted the LSTM [10] deep learning classifier to automatically detect/classify the endangered whale calls. The proposed method was evaluated for real recorded ocean acoustic data from the northeast (NE) Pacific, giving high performances in terms of classification results with a small dataset.

The main contributions of this work are the following: (1) Study of the applicability of the WST to detect/classify endangered whale vocalizations with a small dataset. (2) Incorporation of the temporal contextual information provided by the LSTM network for identification of endangered whale calls. (3) Proposal of an efficient deep-learning-based endangered whale monitoring method from small data samples. (4) To the best of our knowledge, the WST and LSTM techniques together have not been explored for identification of endangered whale calls.

The organization of the paper is as follows. Section 2 presents the materials and methods related to this study. In Section 3, the experimental results are described. The discussion is provided in Section 4.

## 2. Materials and Methods

### 2.1. Dataset

The dataset was constructed using a csv file containing more than 8000 manual annotations for marine mammals (e.g., various whales). Each annotation shows the timestamp (i.e., the start time and the end time), class type and the name of the original wav file from Ocean Networks Canada (ONC)'s database [11]. Each annotated wav file has a duration of 5 min for a sampling frequency of 64 kHz. Prior to doing the annotation, the original files were segmented and cropped into clips. While the same file name can be referred to many of the annotations within the ONC database, the corresponding cropped files were named using a hash of the original csv file's annotation properties. The hashed file name provides the following advantages. First, it gives a unique filename to each annotation. Second, it restricts the processed scripts for downloading and cropping the annotations.

For manual annotation, spectrogram analysis was carried out for recordings every 5 min in Audacity version 2.0.6 [12] software when different whale calls were labeled by an expert from ONC. In total, our generated dataset belonged to 8728 annotated files where files associated with a single class label were considered for use here. In addition, only endangered baleen whales of blue whale and fin whale were taken into account in this study. These data constraints made our dataset reduce to 932 files containing single labels, and the distribution of those files is shown in Table 1. These recordings contain various activities from which we have considered here the activities of the blue whale for 20–100 Hz [2] and the fin whale for 5–100 Hz [5,13].

**Table 1.** The number of sound (wav) files used.

| Endangered Whale | No. of Recordings |
|---|---|
| Blue whale (BW) | 217 |
| Fin whale (FW) | 715 |

### 2.2. Data Analysis

The following stages are taken into account for detection/classification of the endangered whale calls.

#### 2.2.1. Data Preprocessing

Firstly, we reduced the sampling rate of the data segments from 64,000 Hz to 6400 Hz. Secondly, we resized the resampled data segments into data blocks with lengths of $N$ samples, which is set here as $N$ = 64,000 (10 s). To reduce the end effect, each resized block was time-windowed using a $N$-sample Hamming window [14]. It should be noted that computational complexity can be reduced through resampling due to the processing of data at a lower sampling rate. On the other hand, resizing leads to saved memory by compressing the signal along time without modifying its spectral content [15]. Figure 1 shows a flow diagram of the signal processing steps employed in the preprocessing stage. Here, the resampling was performed using a polyphase anti-aliasing filter, whereas the resizing was performed by a "nearest-neighborhood" interpolation.

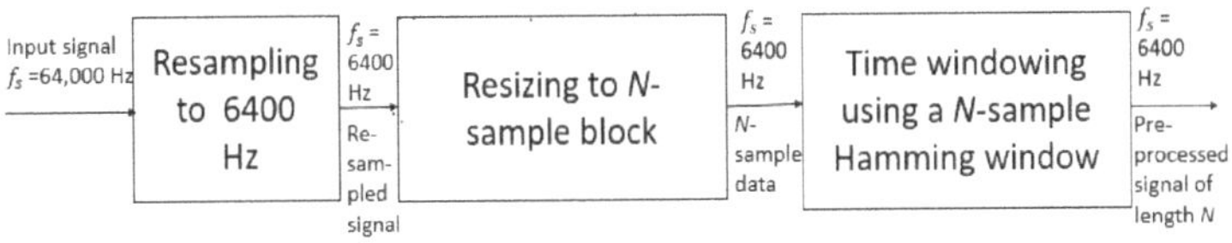


**Figure 1.** Block diagram of the preprocessing steps.

#### 2.2.2. Feature Extraction

The wavelet scattering transform (WST) coefficients are utilized here for feature extraction [16,17]. The 1D WST is computed by cascading wavelet transforms along with nonlinear complex modulus operations followed by average filtering. The WST of a 1D signal $z(t)$ can be represented as

$$S_J z = [S_J^{(0)} z,\ S_J^{(1)} z,\ S_J^{(2)} z] \tag{1}$$

where

$$S_J^{(0)} z(t) = z * \phi_J,$$
$$S_J^{(1)} z(t,\lambda) = |z * \psi_\lambda^{(1)}| * \phi_J, \text{ and}$$
$$S_J^{(2)} z(t,\lambda,\mu) = ||z * \psi_\lambda^{(1)}| * \psi_\mu^{(2)}| * \phi_J$$

In Equation (1), '$*$' denotes the convolution operator, $\psi_\lambda^{(1)}$ and $\psi_\mu^{(2)}$ are the filters representing complex wavelets having center frequencies $\lambda$ and $\mu$, whereas $\phi_J(t)$ is a real lowpass filter with zero-mean frequency.

The implementation of the 1D scattering transform is performed for a given set of wavelet filters whose parameter values are specified initially. Hence, the wavelets are fixed, however, there may be changes for the other parameters after the goal is set, for instance, whether all of $S_J^{(0)} z$, $S_J^{(1)} z$, and $S_J^{(2)} z$, or just $S_J^{(0)} z$ and $S_J^{(1)} z$ would be computed.

While a given input signal length is $N$, the maximum scale of the WST is set to $2^J$. The other issues are the time-frequency resolutions of the wavelets. It is set to $Q = 8$ wavelets per octave for the first-order wavelets, $\psi_\lambda^{(1)}$. On the other hand, it is $Q = 1$, that is, one wavelet per octave as always for the second-order wavelets $\psi_\mu^{(2)}$. These configurations are set to preserve the most signal information for classification.

It is worth noting that postprocessing was performed for the scattering coefficients. Therefore, log-scattering coefficients were obtained by taking the logarithm values for the scattering vectors of the first-order and the second-order wavelets. This can facilitate building the models with reduced dynamic range and stable variability. Moreover, log-scattering coefficients can be well-suited for audio classification due to the fact that amplitudes of the audio signals then vary across several orders of magnitude with no significant changes in the signal content. This process can be characterized through use of the Weber–Fechner law [18] in psychoacoustics.

#### 2.2.3. Classification Method

The classification was performed using Long Short-Term Memory (LSTM) [19,20]. It is a recurrent neural network (RNN) containing an input gate, forget gate, output gate, temporal forward pass and backpropagation. The input gate, forget gate and output gate responses at time $t$ denoted by $i^t$, $o^t$, and $f^t$, respectively, are associated with the forward pass in a LSTM architecture and can be expressed as:

$$i^t = Sigmoid(W_{ih}h^{(t-1)} + W_{ix}x^t + b_i) \tag{2}$$

$$o^t = Sigmoid(W_{oh}h^{(t-1)} + W_{ox}x^t + b_o) \tag{3}$$

$$f^t = Sigmoid(W_{fh}h^{(t-1)} + W_{fx}x^t + b_f) \tag{4}$$

In Equations (2)–(4), $h^{(t-1)}$ refers to the hidden state at time $(t-1)$, $W_{ih}$, $W_{oh}$, and $W_{fh}$ are the weights associated with $h^{(t-1)}$ for the corresponding gates, $b_i$, $b_o$, and $b_f$ are the respective bias vectors, $Sigmoid(x) = \frac{1}{1+e^{(1-x)}}$ is the activation function.

The following formulations are also associated with the forward pass:

$$d^t = Tanh(W_{dh}h^{(t-1)} + W_{dx}x^t + b_d) \tag{5}$$

$$c^t = f^t \odot c^{(t-1)} + i^t \odot d^t \tag{6}$$

$$h^t = o^t \odot Tanh(c^t) \tag{7}$$

$$L^t = \phi(h^t) \tag{8}$$

$$L = \sum_{t=1}^{T} L^t \tag{9}$$

where $d^t$ stands for the distorted input to the memory cell at time $t$, $W_{dh}$ is the weight associated with $h^{(t-1)}$ and $b_d$ is the corresponding bias vector, $Tanh(\cdot)$ is the activation function, $c^t$ refers to the state of the memory cell at time $t$, $h^t$ denotes the hidden state at time $t$, and '$\odot$' stands for point-wise multiplication. Additionally, in Equation (8), $\phi$ maps the hidden state to the network loss $L^t$ at time $t$. Then the total network loss $L$ is found by adding each individual network loss $L^t$ along time, as depicted in Equation (9).

In order to optimize the LSTM model, backpropagation through time was implemented and the most critical value to calculate in LSTM is:

$$\frac{\partial L}{\partial c^t} = \sum_{t=1}^{T} \frac{\partial L^t}{\partial c^t} \tag{10}$$

A critical iterative property was adopted to calculate the above value:

$$\frac{\partial L}{\partial c^{(t-1)}} = \frac{\partial L}{\partial c^t}\frac{\partial c^t}{\partial c^{(t-1)}} + \frac{\partial L^{(t-1)}}{\partial c^{(t-1)}} \tag{11}$$

Several other LSTM gradients can be calculated through the chain rule using the above calculation output:

$$\frac{\partial L}{\partial o^t} = \frac{\partial L}{\partial h^t}\frac{\partial h^t}{\partial o^t}, \tag{12}$$

$$\frac{\partial L}{\partial i^t} = \frac{\partial L}{\partial c^t}\frac{\partial c^t}{\partial i^t}, \tag{13}$$

$$\frac{\partial L}{\partial f^t} = \frac{\partial L}{\partial c^t}\frac{\partial c^t}{\partial f^t}, \tag{14}$$

$$\frac{\partial L}{\partial d^t} = \frac{\partial L}{\partial c^t}\frac{\partial c^t}{\partial d^t} \tag{15}$$

(see [21] for more details).

## 3. Results

The experimental setup and the corresponding detection/classification results of the proposed method, as well as relevant state-of-the-art methods, are presented in the following.

### 3.1. Training Data and Test Data

The dataset used for our results consists of 217 hydrophone recordings for blue whale calls and 715 hydrophone recordings for fin whale calls (sampled at 64,000 Hz) (see Table 1). This dataset includes the noisy signals containing blue/fin whale calls. In order to obtain noise-only recordings, we performed zero-phase filtering for each of the recordings using a fourth-order highpass Butterworth filter with a cut-off frequency of 2000 Hz. Therefore, we have the same number of noise-only recordings. These recordings are then concatenated when three combined datasets containing recordings with and without blue whale calls, with and without fin whale calls, with blue whale and with fin whale calls, are used for our simulations.

For training and testing purposes, each feature set was partitioned into two subsets, namely, the training feature set and test feature set. We used 50% of the data for training and the other 50% for testing in all our simulations. Then the feature sets were standardized with zero-mean and unit variance before input into the classifier. The results were obtained in terms of mean classification results over 100 different trials. For each trial we used different training and test datasets whose configurations changed randomly.

### 3.2. Analytic Result

The spectrograms of the three noisy recordings containing blue whale calls are shown in Figure 2a–c, while the corresponding noise-only signals are presented in Figure 2d–f. Each spectrogram was configured for an input signal of 10 s by using a Hann window of a length of 1600 samples (250 ms) with 75% overlap when the sampling frequency of the signal was downsampled to 6400 Hz.

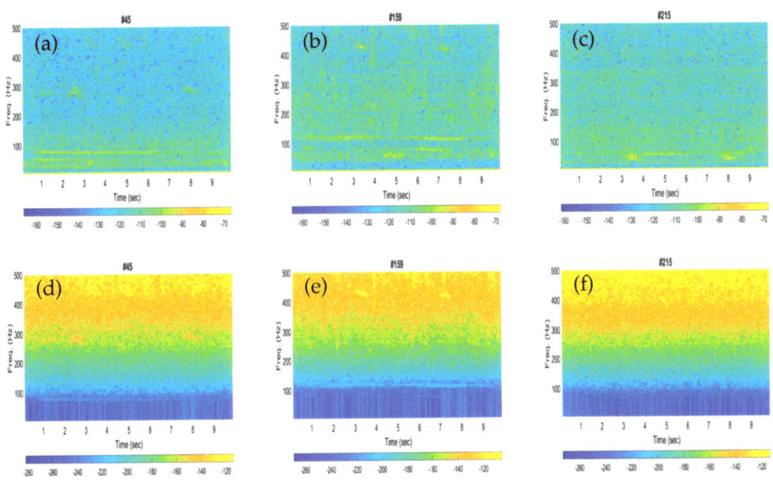


**Figure 2.** Illustrative spectrogram plots for the noisy signals with blue whale calls (**a–c**) and the corresponding noise-only signals (**d–f**).

Similarly, the spectrograms of the three noisy recordings containing fin whale calls are depicted in Figure 3a–c, whereas the corresponding noise-only signals are shown in Figure 3d–f. Each spectrogram was plotted for an input signal of 10 s by using a Hann window of length 1600 samples (250 ms) with 75% overlap, whereas the sampling frequency of the signal was downsampled to 6400 Hz, the same as in Figure 2.

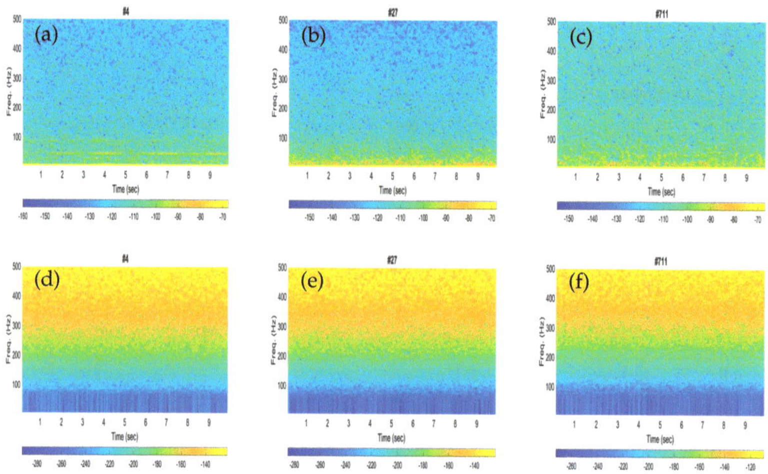

**Figure 3.** Illustrative spectrogram plots for the noisy signals with fin whale calls (**a–c**)) and the corresponding noise-only signals (**d–f**).

We used a three-layer WST and chose Morlet (Gabor) wavelets [22], a commonly used complex wavelets due to its simple mathematical representation and good localization. The framework has two filter banks when the number of layers is three. The quality factors (i.e., the number of wavelet filters per octave) for the first and the second filter banks were set to $Q = 8$ and 1, respectively.

For an input signal of length $N = 64000$ samples and the $Q$ values as above, the output of the framework is a feature matrix with size ($246 \times 8 \times 2$). The feature matrix is then formed with 246 scattering paths and 8 scattering time windows for both the real and imaginary parts of the signal. Hence, the feature set contains 492 feature vectors with dimension 8, while we have excluded the feature vectors associated with path 1. For an $M$ number of signals, a three-dimensional feature output of size ($492 \times 8 \times M$) was thereby generated. In order to build the feature set for the classifier, we multiplied the values of 492 and 8 to reduce them to a 1D sequence and thereby convert the feature output for $M$ signals from three dimensions to two dimensions.

In our training process, we have chosen the following parameters for the LSTM classifier: the number of hidden layers = 512, learning rate = 0.0001, minibatch size = 128, and Adam (Adaptive Moment Estimation) optimizer to train the model. Note that the above parameter setting for the LSTM network provides us good results and the whole process is implemented in MATLAB2022b [23].

The performances of the method are evaluated in terms of classification accuracy (%) as well as sensitivity (%) and specificity (%) as shown below:

$$\text{Accuracy} = \frac{\text{TP+TN}}{(\text{TP+FP}) + (\text{TN+FN})} \tag{16}$$

$$\text{Sensitivity} = \frac{\text{TP}}{\text{TP+FN}} \tag{17}$$

$$\text{Specificity} = \frac{\text{TN}}{\text{TN+FP}} \tag{18}$$

The average accuracy for the blue whale calls (%) over 100 trials is found to be as high as 91.06% using a single epoch in the learning process. The confusion matrix for a trial with accuracy 97.69(%) is shown in Table 2. The rows of this confusion matrix denote the true class labels and the columns represent labels for the predicted class. In the confusion matrix, the diagonal elements refer to the number of correctly classified samples for different class labels, as indicated by the corresponding row/column label. All the non-diagonal elements of the confusion matrix stand for wrongly classified classes, as in Table 2.

**Table 2.** The confusion matrix for the proposed method with the 'Blue whale (BW) + Noise' dataset (the accuracy (%) is indicated in bold font and calculated as a ratio of the sum of diagonal values to the sum of all values × 100).

| | | Predicted class | | |
|---|---|---|---|---|
| | | Blue Whale (BW) | Noise | Sensitivity (%) |
| True class | Blue whale (BW) | 103 | 5 | 95.37 |
| | Noise | 0 | 109 | 100 |
| | Specificity (%) | 100 | 95.61 | 97.69 |

The average accuracy for the fin whale calls (%) over 100 trials is found to be as high as 100% for a single epoch. The confusion matrix for a trial with accuracy 100(%) is shown in Table 3.

**Table 3.** The confusion matrix for the proposed method with the 'Fin whale (FW) + Noise' dataset (the accuracy (%) is indicated in bold font and calculated as a ratio of the sum of diagonal values to the sum of all values × 100).

| | | Predicted class | | |
|---|---|---|---|---|
| | | Fin Whale (FW) | Noise | Sensitivity (%) |
| True class | Fin whale (FW) | 354 | 0 | 100 |
| | Noise | 0 | 361 | 100 |
| | Specificity (%) | 100 | 100 | 100 |

The average accuracy for the classification of blue whale calls and fin whale calls (%) from 100 trials is achieved as 98.40% for a single epoch. The confusion matrix for a trial with accuracy 97.42(%) is displayed in Table 4.

**Table 4.** The confusion matrix for the proposed method with the 'Blue whale (BW) + Fin whale (FW)' dataset (the accuracy (%) is indicated in bold font and calculated as a ratio of the sum of diagonal values to the sum of all values × 100).

| | | Predicted class | | |
|---|---|---|---|---|
| | | Blue Whale (BW) | Fin Whale (FW) | Sensitivity (%) |
| True class | Blue whale (BW) | 112 | 11 | 91.06 |
| | Fin whale (FW) | 1 | 342 | 99.71 |
| | Specificity (%) | 99.12 | 96.88 | 97.42 |

### 3.3. *The Choice of Invariance Scale*

We have considered different invariance scale which is determined as the time support of the lowpass filter $\phi_J(t)$. Figure 4 shows the classification accuracies at different invariance scales of the WST, while we set the invariance scale of 6 (s) as providing the highest classification accuracy (%). The results illustrated in Figure 4 were obtained for the "Blue whale (BW) + Noise" dataset. As we see, the classification performances show some differences with the changes of invariance scale. When the scale is too large, the convolution partly loses the high-frequency information which could cause deterioration of the accuracy at higher scales. On the other hand, when the scale gets small, the convolution removes less noise that might cause a decrease in the performance at lower scales.

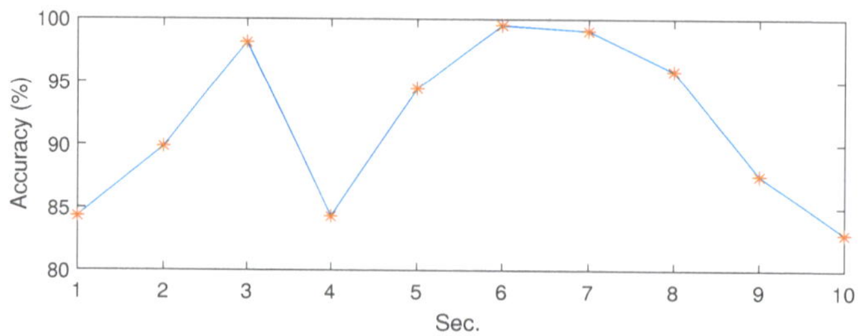


**Figure 4.** The results of classification for various invariance scales.

### *3.4. Results with SVM Classifier*

We have further obtained the results with the support vector machine (SVM) classifier [24] and WST-based features. The SVM classifier basically finds the optimal hyperplane by processing the data using kernels. The optimal hyperplane is produced in terms of the best data separation by maximizing the margin between the decision boundary and the closed data points.

For a given dataset $S = \{(x_i, y_i)\}|x_i \in \mathbb{R}^n, y_i \in \{-1, 1\}_{i=1}^m$, where $n$ refers to the dimensionality of the input data and $m$ is the total number of samples, consider how the expression of a hyperplane is $w \cdot \phi(x) = -b$, where $w$ is the trainable weight vector of the SVM classifier, $x$ is the feature vector, $\phi(\cdot)$ is the kernel function, and $b$ is a bias. Thus, if the point $(x, y)$ is on the hyperplane, $w \cdot \phi(x) + b = 0$; if the point $(x, y)$ is not on the hyperplane, the value of $w \cdot \phi(x) + b$ could be either >0 (positive) or <0 (negative) (considering two classes for binary classification problem optimization). The two SVM parameters, that is, the regularization parameter $C(> 0)$ and the kernel's scale parameter $\gamma$ are set to 10 and 0.05, respectively, where the radial basis function (RBF) is used for the kernel function. The SVM used in our simulations was the least-square SVM (LS-SVM). Additionally, the feature vectors were normalized to zero-mean and unit variance before being fed into the SVM classifier. In Table 5, the average accuracies (%) from 100 trails are presented for the three datasets.

**Table 5.** The average accuracies (%) obtained by using the SVM classifier.

| Dataset | Avg. Accuracy (%) |
|---|---|
| Blue whale (BW) + Noise | 85.34 |
| Fin whale (FW) + Noise | 94.12 |
| Blue whale (BW) + Fin whale (FW) | 100 |

### *3.5. Comparison Result*

#### 3.5.1. Comparison I

We have compared the results with a relevant state-of-the-art method based on a short-term Fourier transform (STFT) and LSTM network [25]. The instantaneous frequency $f_i(t)$ was calculated from the STFT [26] using a 256-sample rectangular window (moving along time with step-size of 1) for feature extraction and used as input to the LSTM classifier. We used an in-built MATLAB function $instfreq(\cdot)$, which estimates the instantaneous frequency to be the first-order spectral moment of the spectrogram, that is, $|\text{STFT}|^2$ of an input signal (see Equation (19)).

$$f_i(t) = \frac{\int_0^\infty f|\text{STFT}(t,f)|^2 df}{\int_0^\infty |\text{STFT}(t,f)|^2 df} \quad (f : \text{frequency}, \ t : \text{time}) \tag{19}$$

In Equation (19), $f_i(t)$ is the instantaneous frequency. The corresponding discrete form of Equation (19) is $f_i(n) = \sum_0^{K-1} k|\text{STFT}(n,k)|^2 / \sum_0^{K-1} |\text{STFT}(n,k)|^2$. Here the discrete form of STFT is $\text{STFT}(n,k) = \sum_{p=-P/2}^{P/2} x(p)g(n-p)e^{-j2\pi kn}$ where $x$ is the signal of length $N$, $g$ is the window of size $(2P+1) \times 1$, $n(1 \leq n \leq N)$ and $k(0 \leq k \leq K-1)$ refer to discrete time and frequency.

The following parameters were used for the LSTM classifier: the number of hidden layers = 100, learning rate = 0.01, minibatch size = 128, and Adam optimizer were used for training the model. Here we have considered the common choices for the values of the LSTM parameters [27].

The average accuracy of the comparison method through 100 trials was found as 69.31% for the blue whale calls. In Table 6, the corresponding confusion matrix of a certain trial is shown where the classification accuracy was obtained as 72.8% with 10 epochs in the learning process.

**Table 6.** The confusion matrix associated with the comparison method I for the 'Blue whale (BW) + Noise' dataset (the accuracy (%) is indicated in bold font and calculated as a ratio of the sum of diagonal values to the sum of all values × 100).

| | | True class | | |
|---|---|---|---|---|
| Predicted class | | Blue whale (BW) | Noise | Specificity (%) |
| | Blue whale (BW) | 89 | 40 | 69.0 |
| | Noise | 19 | 69 | 78.4 |
| | Sensitivity (%) | 82.4 | 63.3 | 72.8 |

Similarly, the mean classification accuracy of this comparison method for the fin whale calls is found as 80.28%. Table 7 presents a confusion matrix associated with the fin whale calls for a particular trial giving 80.4% classification accuracy.

**Table 7.** The confusion matrix associated with the comparison method I for the 'Fin whale (FW) + Noise' dataset (the accuracy (%) is indicated in bold font and calculated as a ratio of the sum of diagonal values to the sum of all values × 100).

| | | True class | | |
|---|---|---|---|---|
| Predicted class | | Fin whale (FW) | Noise | Specificity (%) |
| | Fin whale (FW) | 283 | 69 | 80.4 |
| | Noise | 71 | 292 | 80.4 |
| | Sensitivity (%) | 79.9 | 80.9 | 80.4 |

In addition, we have calculated the mean accuracy for the blue whale and fin whale calls by this comparison method, giving a result of 78.75% for 100 trials. The corresponding confusion matrix for a single trial providing a classification accuracy of 81.1% is displayed in Table 8.

**Table 8.** The confusion matrix associated with the comparison method I for the 'Blue whale (BW) + Fin whale (FW)' dataset (the accuracy (%) is indicated in bold font and calculated as a ratio of the sum of diagonal values to the sum of all values × 100).

| Predicted class | True class | | |
|---|---|---|---|
| | Blue whale (BW) | Fin whale (FW) | Specificity (%) |
| Blue whale (BW) | 51 | 34 | 60 |
| Fin whale (FW) | 54 | 327 | 85.8 |
| Sensitivity (%) | 48.6 | 90.6 | 81.1 |

### 3.5.2. Comparison II

Here, we compared with another relevant state-of-the-art method that is based on scattergram and deep CNN (Convolutional Neural Network) [28]. The scattergram of size $(n \times m)$ was computed using WST and is similar to the mel-spectrogram when considering the filter bank 1 or layer 1 to compute the WST for finding the scattergram. On the other hand, CNN is a popular deep learning approach for classification. The CNN consists of three convolution blocks and one fully connected (FC) layer. Each convolution block is composed of a 1D convolution layer of length 3 and batch normalization. Each convolutional layer is followed by a max pooling layer, with pooling size (1 × 2) and stride (1, 2). The network has 8, 16 and 32 filters, respectively. A fully connected layer with $C$ hidden neurons, where $C$ is the number of classes to be identified, is connected to a categorical softmax layer. We used a rectified linear unit (ReLU) as the activation function in all layers. This architecture takes the scattergram of the fixed-length acoustic data being an input image. The flowchart of the CNN architecture, together with the number of filters and the size of the kernels, are presented in Figure 5.

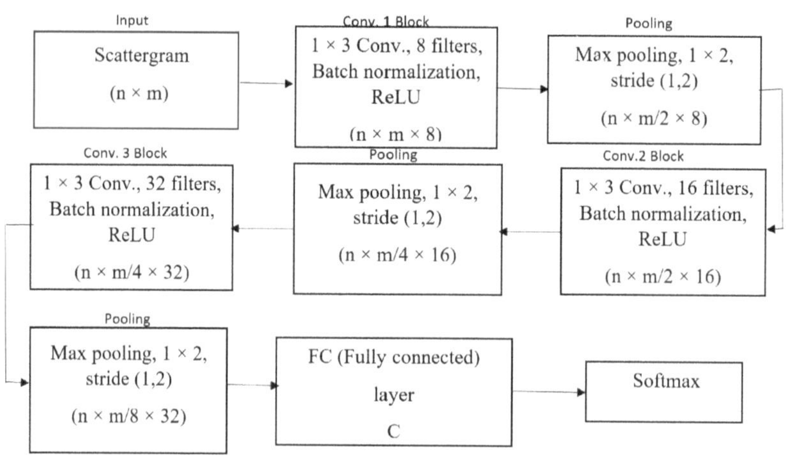


**Figure 5.** The CNN architecture used.

The CNN was trained through a stochastic gradient descent (SGD) optimizer. Similar to the proposed method, a learning rate of 0.0001 and a single epoch were used for the results.

In Table 9, the average classification accuracies (%) for 100 trails and three datasets are presented as obtained by the above comparison method II.

**Table 9.** The average accuracies (%) obtained by comparison method II based on the Scattergram and CNN classifier.

| Dataset | Avg. Accuracy (%) |
|---|---|
| Blue whale (BW) + Noise | 72.26 |
| Fin whale (FW) + Noise | 81.64 |
| Blue whale (BW) + Fin whale (FW) | 48.20 |

## 4. Discussion

Through the LSTM, the temporal context inside the feature set are fully considered and the nonlinear mapping relationship between the past and future information of the signal are learned. The new approach demonstrates significant improvements in the endangered whale calls identification with high classification results as above 90% with small data samples. The method is also computationally efficient, since only a single epoch for the deep learning process is able to produce high classification accuracies (%), as shown in Tables 2–4.

The performances obtained with the fin whales were found to be slightly better than that of the blue whales in terms of higher classification scores. This could be clarified from the spectrogram plots of the illustrative blue whale and fin whale calls. For instance, the fin whale 20 Hz pulses are quite prominent in the spectrograms (see Figure 3b,c), while the blue whale B-calls (40–50 Hz) are less visible in the corresponding spectrogram plots (see Figsure 2a,b). In Tables 2 and 3, the sensitivity represents the correctly classified of the respective whale calls, and the specificity represents the correctly classified of the noise. Then the false positive rates (FPRs) obtained from the specificity (%) were found to be as low as 4.39% and 0%, respectively.

In our proposed framework, the variability is linearized by the WST providing invariants to translations through such average pooling. Most importantly, WST comprising the LSTM network can produce good identification results with small sets of training data. In fact, WST can assist us to extract significant features for LSTM in those situations when it is not possible to learn efficient features with the available training data in case of data scarcity. This makes our proposed scheme for use, such as few-shot learning [29] for identification of endangered whale calls unlike the existing recent deep-learning-based methods [2,3,7,8] that require large datasets for learning.

Both LSTM and CNN are deep neural networks, although the design mechanism of LSTM is different than a CNN. Usually, the LSTM is designed to process and perform prediction/classification from sequences of data by exploiting temporal correlation, whereas the CNN is designed to process image data (for example, a 2D scattergram image shown in Figure 5) for classification by exploiting spatial correlation. The LSTM architecture solved the zero-gradient and exploded gradient, as well as short-term memory problems in RNN. On the other hand, CNN automatically generates complex features at different layers. Basically, WST can be realized as a CNN with fixed filters. In this way, we have a new mechanism here for the classification of endangered whale calls by combining CNN and LSTM. This proposed combined scheme can be efficient, both in terms of classification accuracy and computation.

The proposed method outperforms the relevant state-of-the-art methods based on STFT and LSTM. The presented method provides much-improved classification results even with 1 epoch instead of 10 epochs as used by the comparison method in the learning process. Then the performances of the proposed method is better than the comparison method in terms of classification accuracy as well as computational burden. We have further compared our method with another relevant method based on scattergram and deep CNN, while our method yields better performance. Moreover, the proposed scheme demonstrates high noise resiliency when we compare the results of the proposed scheme and the SVM classifier-based scheme summarized in Table 5.

To the best of our knowledge, the framework consisting of the WST and LSTM networks has not been investigated for detection/classification of whale calls. The preliminary results are presented in this paper. For future work, we would like to obtain data from noisier environments to make the proposed method more robust by proposing a learned wavelet scattering transform together with optimizing the model parameters of the LSTM network. Moreover, we plan to investigate the method for other critically endangered species, such as the North Atlantic right whale and Sei whale.

Finally, we want to emphasize that the MATLAB source codes will be available upon request to the corresponding author.

**Funding:** This research received no external funding.

**Institutional Review Board Statement:** Not applicable.

**Informed Consent Statement:** Not applicable.

**Data Availability Statement:** Not applicable.

**Acknowledgments:** The author gratefully acknowledges the support from Ocean Networks Canada (ONC). The author would like to thank Jasper Kanes at ONC for giving her expert annotations of the dataset as well as her valuable insights about the annotations. The author also would like thank the reviewers and the associate editor for their valuable comments and suggestions that helped to improve the paper.

**Conflicts of Interest:** The author declares no conflict of interest.

## References

1. Koubrak, O.; VanderZwaag, D.L.; Worm, B. Endangered blue whale survival in the North Atlantic: Lagging scientific and governance responses, charting future courses. *Int. J. Mar. Coast. Law* **2022**, *37*, 89–136. [CrossRef]
2. Zhong, M.; Torterotot, M.; Branch, T.A.; Stafford, K.M.; Royer, J.-Y.; Dodhia, R.; Ferres, J.L. Detecting, classifying, and counting blue whale calls with Siamese neural networks. *J. Acoust. Soc. Am.* **2021**, *149*, 3086–3094. [CrossRef] [PubMed]
3. Miller, B.S.; Madhusudhana, S.; Aulich, M.G.; Kelly, N. Deep learning algorithm outperforms experienced human observer at detection of blue whale D-calls: A double-observer analysis. *Remote Sens. Ecol. Conserv.* **2023**, *9*, 104–116. [CrossRef]
4. Cooke, J.G. Balaenoptera physalus. In *IUCN Red List of Threatened Species*; IUCN Report T2478A50349982; International Union for Conservation of Nature: Cambridge, UK, 2018. [CrossRef]
5. Aulich, G.M.; McCauley, R.D.; Saunders, B.J.; Parsons, M.J. Fin whale (*Balaenoptera physalus*) migration in australian waters using passive acoustic monitoring. *Sci. Rep.* **2019**, *9*, 8840. [CrossRef] [PubMed]
6. Burkhardt, E.; van Opzeeland, I.; Cisewski, B.; Mattmuller, R.; Meister, M.; Schall, E.; Spiesecke, S.; Thomisch, K.; Zwicker, S.; Boebel, O. Seasonal and diel cycles of fin whale acoustic occurance near elephant island, Antarctica. *R. Soc. Open Sci.* **2021**, *8*, 201142. https://org/doi/10.1098/rsos.201142. [CrossRef] [PubMed]
7. Rasmussen, J.H.; Sirovic, A. Automatic detection and classification of baleen whale social calls using convolutional neural networks. *J. Acoust. Soc. Am.* **2021**, *149*, 5. [CrossRef] [PubMed]
8. Garibbo, S.; Blondel, P.; Heald, G.; Heyburn, R.; Hunter, A.J.; Williams, D. Characterising and detecting fin whale calls using deep learning at the Lofoten-Vesterålen Observatory, Norway. *Proc. Meet. Acoust.* **2021**, *44*, 070021. [CrossRef]
9. Shamaei, A.; Starcuková, J.; Star, Z., Jr. A wavelet scattering convolutional network for magnetic resonance spectroscopy signal quantitation. In Proceedings of the 14th International Joint Conference on Biomedical Engineering Systems and Technologies (BIOSTEC'2021), Vienna, Austria, 11–13 February 2021.
10. Juneau, P.; Baddour, N.; Burger, H.; Bavec, A.; Lemaire, E.D. Comparison of decision tree and long short-term memory approaches for automated foot strike detection in lower extremity amputee populations. *Sensors* **2021**, *21*, 6974. [CrossRef] [PubMed]
11. Ocean Networks Canada. Ocean Networks Canada Data Archive: Hydrophone Data from 1 January 2014 to 4 December 2014. Available online: http://www.oceannetworks.ca (accessed on 3 May 2020).
12. Available online: http://audacity.sourceforge.net/ (accessed on 30 June 2020).
13. Ford, J.K.B.; Koot, B.; Vagle, S.; Patch, N.H.; Kamitakahara, G. Passive acoustic monitoring of large whales in offshore waters of British Columbia. *Can. Tech. Rep. Fish. Aquat. Sci.* **2010**, *2898*, 30. Available online: https://www.mals/all-about/killer-whale/communication/.com (accessed on 1 August 2022).
14. Oppenheim, A.V.; Schafer, R.W. *Digital Signal Processing*; Prentice-Hall: Englewood Cliffe, NJ, USA, 1975.
15. Proakis, J.; Manolakis, D. *Digital Signal Processing: Principles, Algorithms and Applications*; Macmillan Publishing Company: New York, NY, USA, 1992.
16. Bruni, V.; Cardinali, M.L.; Vitulano, D. An MDL-based wavelet scattering features selection for signal classification. *Axioms* **2022**, *11*, 376. [CrossRef]

17. Marzog, H.A.; Abd, H.J. Machine learning ECG classification using wavelet scattering of feature extraction. *Appl. Comput. Intell. Soft Comput.* **2022**, *2022*, 9884076. [CrossRef]
18. Norwich, K.H.; Wong, W. Unification of psychophysical phenomena: The complete form of Fechner's law. *Percept. Psychophys.* **1997**, *59*, 929–940. [CrossRef] [PubMed]
19. Hochreiter, S.; Schmidhuber, J. Long short-term memory. *Neural Comput.* **1997**, *9*, 1735–1780. [CrossRef] [PubMed]
20. Houdt, G.V.; Mosquera, C.; Napoles, G. A review on the long short-term memory model. *Artif. Intell. Rev.* **2020**, *53*, 5929–5955. [CrossRef]
21. Rahuljha. LSTM Gradients. Towards Data Science, 2020. Available online: https://towardsdatascience.com/lstm-gradients-b3996e6a0296 (accessed on 1 July 2022).
22. Chen, H.; Zuo, M.J.; Wang, X.; Hoseini, M.R. An adaptive Morlet wavelet filter for time-of-flight estimation in ultrasonic damage assessment. *Measurement* **2010**, *43*, 570–585. [CrossRef]
23. MATLAB. *Version R2022b*; The MathWorks Inc.: Natick, MA, USA, 2022.
24. Sattar, F.; Driessen, P.F.; Tzanetakis, G.; Page, W.H. A new event detection method for noisy hydrophone data. *Appl. Acoust.* **2020**, *159*, 107056. [CrossRef]
25. Wang, T.; Kirchner, F. Grasp stability prediction with time series data based on STFT and LSTM. *arXiv* **2021**, arXiv:2106.01272.
26. Boashash, B. Estimating and interpreting the instantaneous frequency of a signal. II. algorithms and applications. *Proc. IEEE* **1992**, *80*, 540–568. [CrossRef]
27. Reimers, N.; Gurevych, I. Optimal hyperparameters for deep LSTM-networks for sequence labeling tasks. *arXiv* **2017**, arXiv:1707.06799.
28. Tanveer, M.H.; Zhu, H.; Ahmed, W.; Thomas, A.; Imran, B.M.; Salman, M. Mel-Spectrogram and Deep CNN Based Representation Learning from Bio-Sonar Implementation on UAVs. In Proceedings of the IEEE International Conference on Computer, Control and Robotics (ICCCR), Shanghai, China, 8–10 January 2021. [CrossRef]
29. Yang, J.; Guo, X.; Li, Y.; Marinello, F.; Ercisli, S.; Zhang, Z. A survey of few-shot learning in smart agriculture: Developments, applications, and challenges. *Plant Methods* **2022**, *18*, 28. [CrossRef] [PubMed]

*sensors*

MDPI

*Article*

# Microstructure Turbulence Measurement in the Northern South China Sea from a Long-Range Hybrid AUV

**Yunli Nie [1], Xin Luan [2], Yan Huang [3], Libin Du [1,*], Dalei Song [4,5,*] and Xiuyan Liu [6]**

1 College of Ocean Science and Engineering, Shandong University of Science and Technology, Qingdao 266590, China
2 College of Information Science and Engineering, Ocean University of China, Qingdao 266100, China
3 Shenyang Institute of Automation Chinese Academy of Sciences, Shenyang 110016, China
4 College of Engineering, Ocean University of China, No. 238 Songling Rd, Qingdao 266100, China
5 Institute for Advanced Ocean Study, Ocean University of China, Qingdao 266100, China
6 School of Information and Control Engineering, Qingdao University of Technology, Qingdao 266100, China
* Correspondence: dulibin@sdust.edu.cn (L.D.); songdalei@ouc.edu.cn (D.S.)

**Abstract:** This study describes the development of a long-range hybrid autonomous underwater vehicle (AUV) for ocean turbulence measurement. It is a unique instrument, combining the characteristics of the conventional AUV and the buoyancy-driven glider, with a variety of flexible motion modes, such as cruise mode, glider mode, drift mode, and combination of multiple motion modes. The hybrid AUV was used for continuous turbulence measurement in the continental slope of the northern South China Sea in 2020. A total of ten continuous profiles were completed covering a horizontal span of 25 Km and a depth of 200 m. The hybrid AUV was operated in the combined glider and cruise mode. The hybrid AUV's flight performance was stable and satisfied the requirement for turbulence observation. The measured velocity shears from both probes were in good agreement, and the noise-reduced shear spectra were in excellent agreement with the Nasmyth spectrum. The water column in the study area was highly stratified, with a thick thermocline. The dissipation rate ($\varepsilon$) varied from $1.41 \times 10^{-10}$ to $4.18 \times 10^{-7}$ W·kg$^{-1}$. In the surface mixed layer, high values of $\varepsilon$ ($10^{-9}$~$10^{-8}$ W·kg$^{-1}$) were observed toward the water surface. In the thermocline, $\varepsilon$ was $10^{-9.5}$~$10^{-9}$ W·kg$^{-1}$, which was smaller than the level of the surface mixed layer. This result was mainly because of the strong "barrier"-like thermocline, which damped the transmission of wind and heat energy from the surface mixed layer to the deep layer. Overall, this study demonstrates the utility of hybrid AUVs for collecting oceanic turbulence measurements. They are a powerful addition to traditional turbulence instruments, as they make it possible to survey large areas to obtain high-quality and high-resolution data in both vertical and horizontal directions over long durations.

**Keywords:** ocean turbulence; hybrid AUV; dissipation rate; thermocline; mixing

## 1. Introduction

In the spread of contaminants, sedimentation processes, and nutrient levels across all ocean zones, turbulent mixing and the resulting dissipation of energy are significant [1,2]. Consequently, it is essential to comprehend the distribution of turbulent energy under various background conditions.

Direct measurements of mixing require high-frequency measurements of variables such as current shear or temperature. According to Lueck et al. (2002) [3], a wide range of instruments for measuring vertical and horizontal turbulence have been developed. For instance, the vertical profiler, which is the instrument that is used the most, is a very quiet platform that does not have any mechanical vibration. The application of these instruments demonstrates the turbulence structure's temporal and spatial variability. Vertical profilers, however, are limited in their ability to offer horizontal sampling because of the logistics involved in their deployment, especially in the upper ocean where horizontal inhomogeneity and the impact of phenomena such as Langmuir cells inside the mixed layer can be considerable. A range of instrument, including towed vehicles, submarines, AUVs and free-fall gliders, were used to measure horizontal turbulence beginning in the 1950s [4–7]. Because of their high deployment cost, early horizontal turbulence instruments were not widely used.

Recently, several small AUVs of 2–4 m length with shear probes have been developed to measure ocean turbulence, including the REMUS AUV, the glider, and others [8–11]. Their dissipation rates can be as low as $10^{-11}$ $W{\cdot}kg^{-1}$. The REMUS AUV has flexible mission types capable of accomplishing horizontal and vertical as well as bottom and surface observation tasks. It has been shown to outperform vertical profilers in investigating ocean mixed layers in the ocean under various conditions. As a result, using AUVs as a platform for measuring turbulence can result in long-term, continuous measurements in four dimensions. The measured data will help researchers investigate theories of turbulent cascade and stationarity in the oceans as well as comprehend the temporal and spatial variability of turbulent processes.

In this paper, a long-range hybrid AUV was developed to measure ocean turbulence in the northern South China Sea (*nSCS*). The hybrid AUV was a free-swimming, unmanned underwater vehicle that combined the attributes of the conventional AUV and the buoyancy-driven glider. This has made it possible to perform turbulent microstructural and meso-structural measurements that were previously not possible due to time, cost, or platform limitations. Multipass missions at various water depths in deep or shallow water enabled measurements in a variety of conditions within 1 m of the seafloor or sea surface. The methods are described in Section 2, with an emphasis on the design of the long-range hybrid AUV and the turbulence package. Section 3 describes the experiment, including the location and the sampling strategy. Subsequently, the results are presented in Section 4, including hybrid AUV flight performance, as well as local hydrography and microstructure data. Section 5 discusses the turbulent mixing characteristics of the *nSCS* and its relationship with the thermocline. Section 6 provides a brief conclusion of the study.

## 2. Methods

### *2.1. The Long-Range Hybrid AUV*

The long-range hybrid AUV is shown in Figure 1. It has a length of 3 m, a diameter (D) of 0.35 m and a mass of 200 kg [12,13]. For ocean surveys, the system has an endurance of 1500 km and can dive to 2000 m for oceanographic research. In terms of cruising capability, the vehicle is designed to reach up to 1.5 m/s at full speed. To execute different observation tasks with high efficiency, the design of the hybrid AUV combines the features of the glider and the conventional AUV. The buoyancy-driven cabin primarily features a fixed battery pack and a buoyancy control system that varies the net buoyancy of the system by adjusting the volume of the external oil bladder. The attitude-regulating cabin is designed to move the battery pack forward or backward to adjust the vehicle's center of gravity. The propeller propulsion unit, the elevators, and the rudders make up the aft cabin. The elevators and the two rudders are coaxial. Therefore, the long-range hybrid AUV has a variety of flexible motion modes, such as yo-yo profiles similar to gliders (glider mode), cruising at a required depth (cruise mode), drifting with very little power consumption (drift mode), and combination of multiple motion modes, as shown in Figure 2.

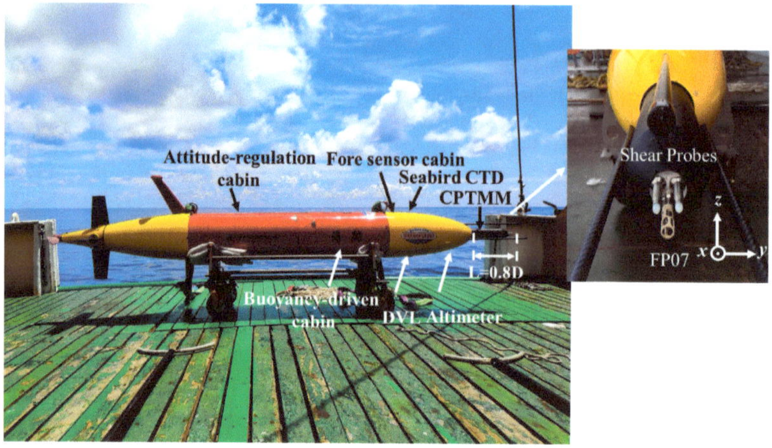


**Figure 1.** The structural components and oceanographic sensors of the long-range hybrid AUV. It is 3 m long, 0.35 m diameter, 200 kg mass and has a 1500 km endurance. Shown are the various micro and fine structure sensor systems. These include the Seabird GPCTD, the Nortek DVL, the dissolved oxygen, and the CPMTM for turbulence measurements. The CPMTM is in the middle of the fore sensor cabin and the shear probes are 0.8 D away from the nose.

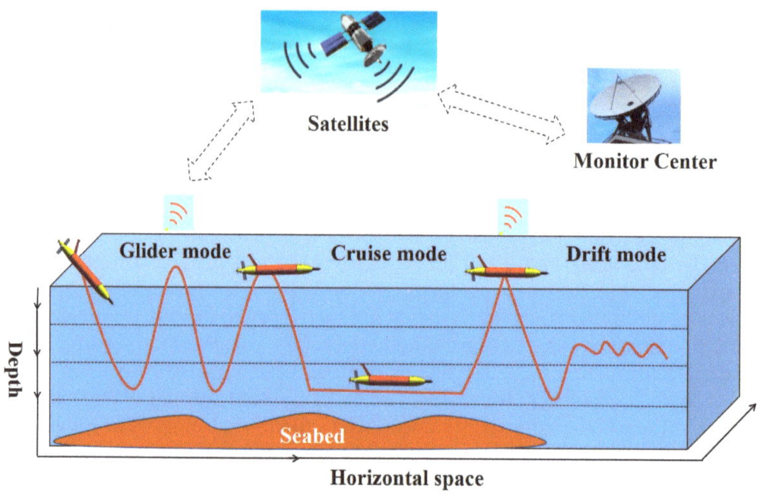


**Figure 2.** Typical working modes of the long-range hybrid AUV are: the glider mode, the cruise mode, and the drift mode. The long-range hybrid AUV combines the attributes of AUV and buoyancy-driven glider and has excellent maneuverability.

A cross-platform instrument for microstructure turbulence measurements (CPMTM), a Seabird conductivity, temperature, and depth (CTD) sensor (with dissolved oxygen), a Doppler velocity log (DVL), and an altimeter were mounted on the hybrid AUV. All sensors were in the fore sensor cabin, which was near the bow (Figure 1). The low drag profile of the hybrid AUV was maintained while the data sampling accuracy was improved by separating the sensors from the rear propulsion system. Table 1 shows the performance parameters of the major sensors of the hybrid AUV. This array of sensors in a hybrid AUV enabled the quantification of important dynamic and kinematic turbulence and microscopic physical processes.

**Table 1.** The hybrid AUV measured parameters, sensor type, and sensor specifications.

| Parameter | Range | Precision | Resolution | Fs (Hz) |
|---|---|---|---|---|
| Current profiling | 16 m/s | ±0.1%/±0.1 cm/s | 0.1 cm/s | 8 |
| Temperature | −5–42 °C | ±0.002 °C | 0.001 °C | |
| Conductivity | 0–9 S/m | ±0.0003 S/m | 0.00001 S/m | 1 |
| Depth | 2000 m | ±0.1%/FS | 0.002%/FS | |
| Dissolved oxygen | 120% of saturation | ±2% of saturation | / | 1 |
| Velocity shear | 0–10 $s^{-1}$ | 5% | 10–3 $s^{-1}$ | 1024 |
| Fast temperature | −5–35 °C | 0.005 °C | 10–5 °C | 1024 |
| Accelerometer | ±2 g | ±1% | 10–5 g | 512 |

### 2.2. CPMTM

The CPMTM was an "all-in-one" payload with a length of approximately 0.6 m and a diameter of 0.08 m, and it had a flexible vibration-damping device inside it [14]. The performance of the instrument and the turbulent flow characteristics were both measured by the CPMTM's numerous sensors. The cross-stream velocity variations (∂y/∂x) and the vertical velocity gradient (∂z/∂x) were measured using the shear probes [15], which were positioned orthogonally. The dissipation rate's effective noise floor was discovered to be less than $10^{-11}$ W·$kg^{-1}$ [14]. Microstructure temperature fluctuations were measured with a fast thermistor (FP07) [16], with a response time of approximately 7 ms. The CPMTM was fitted with a highly sensitive 3-axis accelerometer to measure the intensity of vibration during profiling. Shear signals could be post-processed to eliminate body vibrations. All turbulence channels (temperature and shear) were sampled at 1024 Hz, whereas the accelerometer channel was sampled at 512 Hz. The CPMTM was in the middle of the fore sensor cabin and the shear probes are 0.8 D away from the nose (Figure 1).

## 3. Experiment

The South China Sea is the marginal sea of the Western Pacific Ocean, which is mainly composed of shoals, continental shelves, and deep-sea basins. Through the Luzon Strait, it is connected to the Western Pacific Ocean in its northeastern direction. The continental shelf break zone of the *nSCS* is very wide, and the terrain changes dramatically. Within fewer than 10 nautical miles, the water depth rapidly changes from 100 m to 1000 m. It is precisely because of the interaction between topography and tide that the activities of internal waves and internal tides in this region are very active and dissipated here, resulting in strong turbulent mixing [17].

From September 10th to 17th, 2020, A continuous and large-range turbulence measurement was performed on the slope of the *nSCS* aboard R/V *Yuezhanyuke 8*. Figure 3a depicts the measurement site within the *nSCS*, where the water depth was approximately 2000 m. Figure 3b displays the hybrid AUV's precise track. From 06:00 to 17:00 on 14 September 2020 (local time), the hybrid AUV was observed along the 18°30′ N section from east to west. The operating mode was a combination of glider and cruise modes, with the glider mode having a pitch angle of about 13° between the surface and 200 m depth, and the cruise mode lasting 5 min at 200 m depth. Several high-quality spatiotemporal turbulence data of the upper ocean were collected as a result of the measurements, which produced a total of ten continuous profiles with a horizontal distance of 25 km. In the experiment, the hybrid AUV was used as the platform for our measurement, as shown in Figure 4.

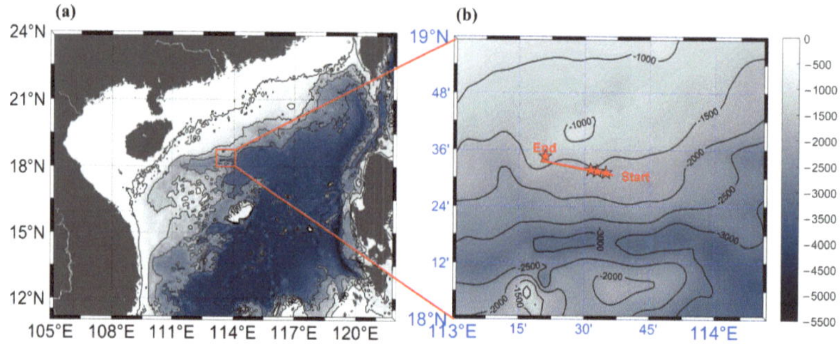


**Figure 3.** (**a**) Topography of the *nSCS* and the measuring area. (**b**) A larger image of the region enclosed by the red rectangle in panel (**a**), displaying the hybrid AUV path's start and end points as well as the surrounding bathymetry. Color in both panels denotes the depth (m) of the water.

**Figure 4.** Deployment of the hybrid AUV from *R/V Yuezhanyuke* 8: (**a**) the hybrid AUV is ready on the deck, (**b**) the hybrid AUV is attached to the cables and lowered into the water through the A-frame of the *R/V*, (**c**) the hybrid AUV is released and starts work, and (**d**) the hybrid AUV back to the surface and retrieved.

## 4. Results

### *4.1. Flight Performance*

The direction of the moving hybrid AUV relative to the horizontal is called the glide angle ($\gamma$), it is defined as the pitch angle ($\theta$) plus the "angle of attack" (AOA or $\alpha$):

$$\gamma = \theta + \alpha, \quad (1)$$

The pitch angle ($\theta$) is measured by the hybrid AUV's attitude sensor, but the AOA is not measured directly. For high-quality dissipation rate calculation, crucial parameters such as the hybrid AUV speed through water ($U$) and the AOA need to be estimated as accurately as possible [18]. The AOA is obtained here using a hydrodynamic flight model developed by Merckelbach et al. (2010) [19], and uses measured pitch angle and pressure ($P$) to dynamically estimate $U$,

$$U = \frac{W}{sin(\gamma)} = \frac{\partial P/\partial t}{sin(\theta + \alpha)}, \quad (2)$$

where $W = \partial P/\partial t$ is the AUV's vertical speed based on the rate of pressure change.

Figure 5 shows an overview of the data sampled by the hybrid AUV in the course of the mission. This includes time series for depth, heading, roll, pitch, vertical velocity, and $U$. In this mission, 10 continuous profiles were completed and the cruise mode of about 5 min was carried out at a depth of 200 m (Figure 5a). The flight path (Figure 5b) was a straight line with a constant heading of 270°. Figure 5c shows that the roll angle remained constant between 0° and 2°. During descent and ascent, the mean pitch angle was −13.42°/12.78° and the standard deviation was 1.31°/1.24° (Figure 5d). Near the interface, the roll and pitch variance substantially increased, the data of the upper 10 m was deleted. We were able to estimate the hybrid AUV's speed along its flight path using the pitch angle and vertical speed (Figure 5e). Figure 5f shows that during descent/ascent, the mean speed of the hybrid AUV was 0.57 m·s$^{-1}$/0.65 m·s$^{-1}$, and in cruise mode was 0.68 m·s$^{-1}$. The mean and standard deviation values of the hybrid AUV's flight characteristics are summarized in Table 2. The hybrid AUV's flight performance met the requirements for turbulence measurement and was stable.

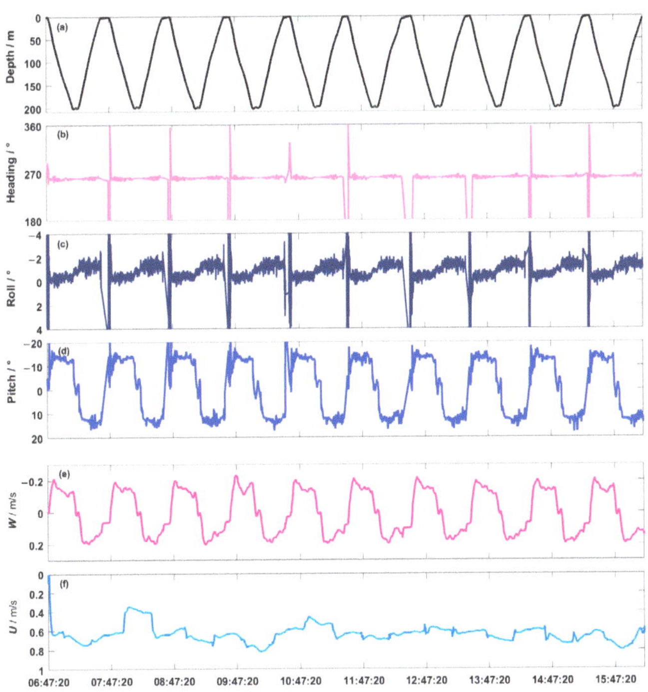


**Figure 5.** Flight performance of the hybrid AUV during the mission includes time series for (**a**) depth, (**b**) heading, (**c**) roll, (**d**) pitch, (**e**) $W$, and (**f**) $U$. The depth record shows the typical pattern with descent, cruise at 200 m, and ascent.

**Table 2.** Mean ± 1 standard deviation of the hybrid AUV flight parameters of the mission (Note. $\theta$ is the pitch angle, $W$ is the vertical speed, and $U$ is the speed through the water).

| | *Pitch* (°) | *Vertical Speed* ($m{\cdot}s^{-1}$) | *U* ($m{\cdot}s^{-1}$) |
|---|---|---|---|
| Descent | −13.42 ± 1.31 | −0.15 ± 0.02 | 0.57 ± 0.07 |
| Ascent | 12.78 ± 1.24 | 0.16 ± 0.02 | 0.65 ± 0.07 |

### 4.2. Hydrography

Weather during the experiment was dominated by sunny conditions. The contour plots of temperature, salinity, and potential density along the section were depicted in Figure 6. The temperature's vertical structure is depicted in Figure 6a. The sea surface temperature of the *nSCS* was about 30 °C, and the overall vertical temperature ranged from 14.9 °C to 30.7 °C. A prominent thermocline (40 m~150 m) and a thin surface mixed layer (~40 m) characterized the water column. The surface mixed layer is defined as the depth at which the temperature changes by 0.5 °C from the surface temperature [20]. In the thermocline, the temperature gradient is greater than 0.1 °C/m, and the temperature variation reaches $\Delta T \approx 10$ °C.

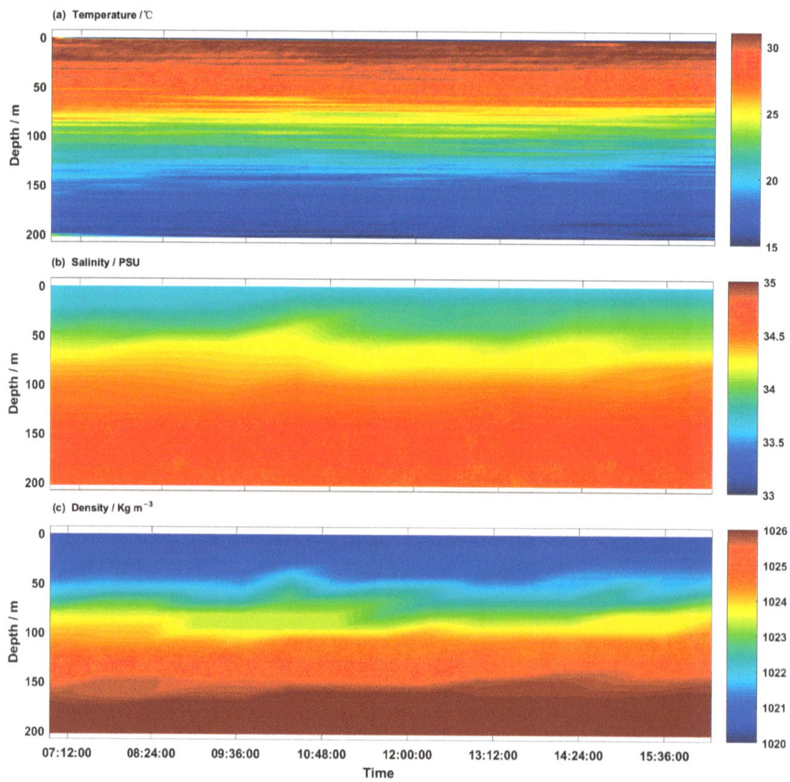


**Figure 6.** Contour plots of (**a**) temperature, (**b**) salinity, and (**c**) potential density measured by the hybrid AUV. According to the data, there was a thermocline between 40 and 150 m thick that separated the surface mixed layer and bottom mixed layer throughout the deployment.

Salinity was 33.5 PSU at the surface and monotonically increased to almost 34.7 PSU below 150 m (Figure 6b). The density structure (Figure 6c) showed that throughout the deployment period, this region was strongly vertically stratified, with an average density difference of 5.9 kg·m$^{-3}$ between the bottom and surface mixed layers. Figure 6a shows that the density structure was mainly controlled by temperature.

### 4.3. Microstructure Data

#### 4.3.1. Data Screening: Shear Probes

The shear probe voltage output was converted to velocity shear using the known shear probe sensitivity, electronic constant, flow through the sensor [3]. When the AUV maneuvers at the turn points, the angle of attack is large and the flow past the shear probe is almost zero. So, the velocity shear data is not available. In addition, when the AUV speed $U > 0.4$ m/s, the velocity shear data are valid.

Therefore, firstly, data screening was carried out according to the AUV's flight performance. Then, according to Rayda's criterion, the singular data of the velocity shear signal were removed and replaced by the arithmetic mean value when the measurement error was three times the standard error [12]. Finally, the velocity shear signal was bandpass filtered from 0.15 Hz to 100 Hz to effectively remove the low-frequency motion and high-frequency vibration signature of the AUV. However, the elimination of this signal does not affect the dissipation rate calculation. A sample of velocity shear, collected during a steady descent of the fifth profile, is shown in Figure 7. Data in the surface were removed due to contamination by the flight performance of the AUV. After the start of the experiment, the velocity shear varied from $-0.5$ s$^{-1}$ to 0.5 s$^{-1}$, and they were very consistent.

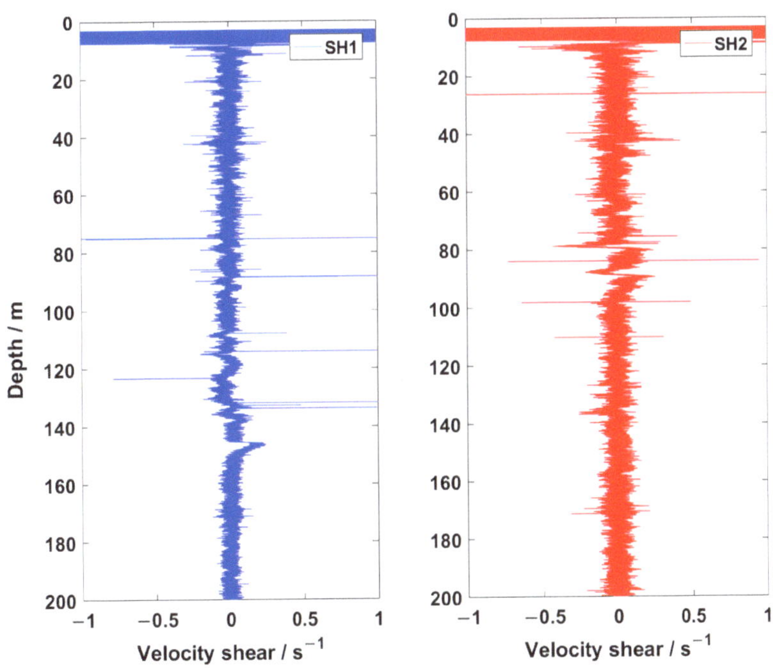


**Figure 7.** Sample of velocity shear of the two shear probes (SH1, SH2), collected during a steady descent of the fifth profile.

#### 4.3.2. Shear Spectra

For spectral analysis, the velocity shear was divided into 12-s-long segment that were half-overlapping. We selected a fast Fourier-transform (FFT) length, corresponding to 4 s, detrend and Hanning window of each 4 s segment prior to calculating the spectra. The acceleration coherent noise was removed from the velocity shear signal using the method proposed by Goodman et al. [21] to reduce contamination from vehicle motion and vibration. For the best results and statistical significance when using the Goodman method, a record period that is longer than the FFT period is recommended. This method is based on the cross spectra between the shear probe and the accelerometer. We applied 12 s segments, which is three times the period of an FFT. The dissipation rate was calculated using the cleaned shear spectrum. Using Taylor's frozen turbulence hypothesis and the $U$, the shear spectra $F(f)$ in the frequency domain was transformed to the wavenumber domain ($k$), where $k = f/U$ and $F(k) = UF(f)$.

Figure 8 shows the wavenumber spectra of the velocity shear in Figure 7 at different depths. It is clear that both shear probes' measured wavenumber spectra (red and green lines) are in agreement with one another. They also match the corresponding Nasmyth spectrum well. These results demonstrate the high quality of the data and the hybrid AUV's capacity to measure ocean turbulence.

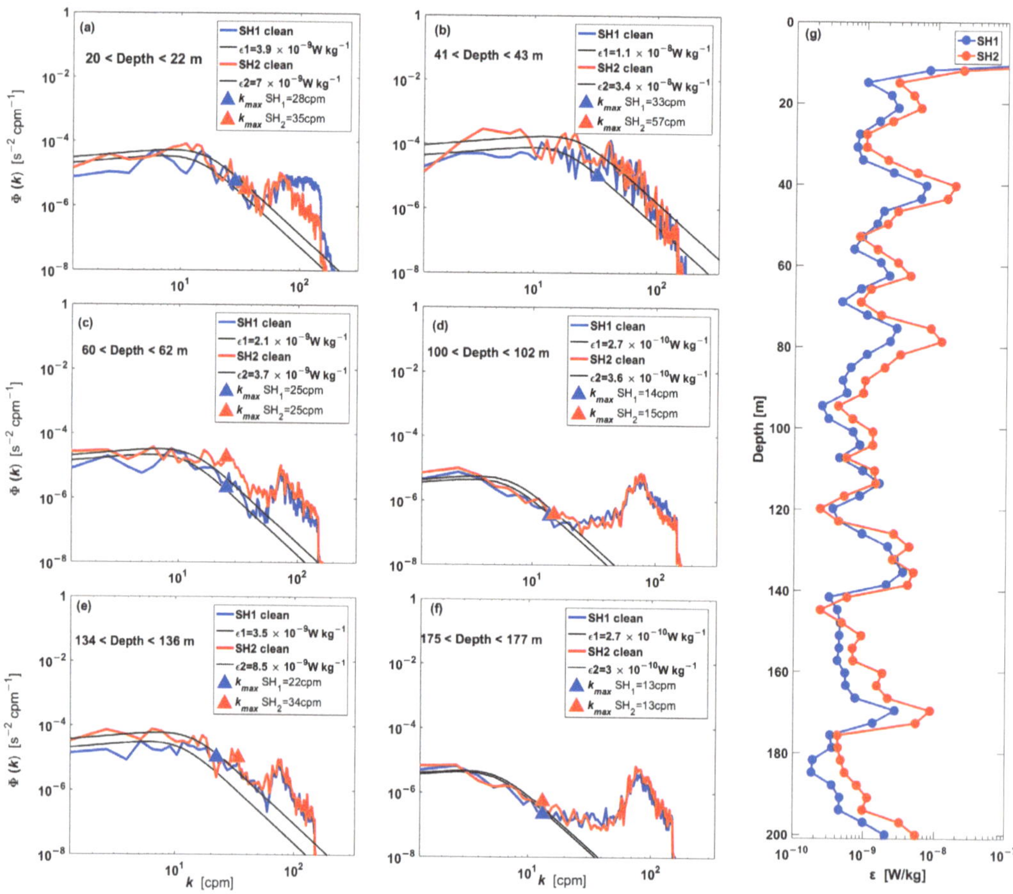


**Figure 8.** (**a–f**) Wavenumber shear spectrum samples from two shear probes (red and blue lines). The Nasmyth spectrum (black line) is shown for each case. Triangles (red and blue) mark the limits of $k_{max}$. The corresponding ε and depth are included in each plot. (**g**) The profile of ε calculated from the velocity shear.

#### 4.3.3. Estimation of Kinetic Energy Dissipation Rate

Assuming isotropic turbulence, the dissipation rate ($\varepsilon$) for each data set is calculated by integrating the wavenumber spectra as follows:

$$\varepsilon_i = 7.5v\overline{\left(\frac{\partial u_i}{\partial x}\right)^2} = 7.5v\int_{k_{min}}^{k_{max}} \Phi_{u_i}(k)dk \tag{3}$$

where $v$ represents the kinematic viscosity of seawater ($\approx 1\times10^{-6}$ $m^2\cdot s^{-1}$), $i$ (= 1, 2) is the number of shear probe, the overbar represents averaging, the $\Phi_{u_i}(k)$ is the estimated spectrum of velocity shear. The lower ($k_{min}$) and upper ($k_{max}$) integration limits of the spectrum are determined using Nasmyth spectrum for the turbulence [22], and the variance in the spectrum's unresolved parts is corrected. We adopt the precise curve fit for the Nasmyth spectrum provided by Wolk et al. [23], when the shear spectrum is higher (lower) the Nasmyth spectrum, $k_{max}$ increases (decreases). According to the Nasmyth spectrum, integrating to $0.5k_k$ where $k_k = (2\pi)^{-1}(\varepsilon/v^3)^{1/4}$ is the Kolmogorov wavenumber, resolves 90% of the variance.

Figure 8 shows some examples of wavenumber spectra and the corresponding estimates of $\varepsilon$. The shear spectra are depicted by the red curves, the Nasmyth spectrum fitted to the observed data are depicted by the black curves, and the $k_{max}$ limits are depicted by the green triangles. Figure 8g shows the profile of $\varepsilon$ calculated from the velocity shear. The average value of $\varepsilon$ was $\sim10^{-9}$ $W\cdot kg^{-1}$.

## 5. Discussion

Figure 9a shows the depth–time map of $\varepsilon$. Due to contamination from the hybrid AUV tilt, the data of the upper 10 m was deleted. The turbulence characteristics in this region are significantly different from those in shallow coastal waters, where turbulent mixing is strongly influenced by the thermocline. In the surface mixed layer, higher values of $\varepsilon$ ($10^{-9}\sim10^{-8}$ $W\cdot kg^{-1}$) were observed towards the water surface, mainly explained by the wind forcing. Variable dissipation was observed in the thermocline, with increased dissipation occurring at the upper boundary of the thermocline. The thermocline time-averaged $\varepsilon$ ranges from $10^{-9.5}$ $W\cdot kg^{-1}$ to $10^{-9}$ $W\cdot kg^{-1}$ (Figure 9b), slightly less than the level of the open ocean thermocline ($10^{-9}$ $W\cdot kg^{-1}$) [24,25]. Compared to the dissipation over the thermocline, the dissipation below it was weak, with time-averaged $\varepsilon$ between $10^{-10}$ $W\cdot kg^{-1}$ and $10^{-9.5}$ $W\cdot kg^{-1}$ (Figure 9b). Dissipation rates were enhanced at the bottom boundary layer.

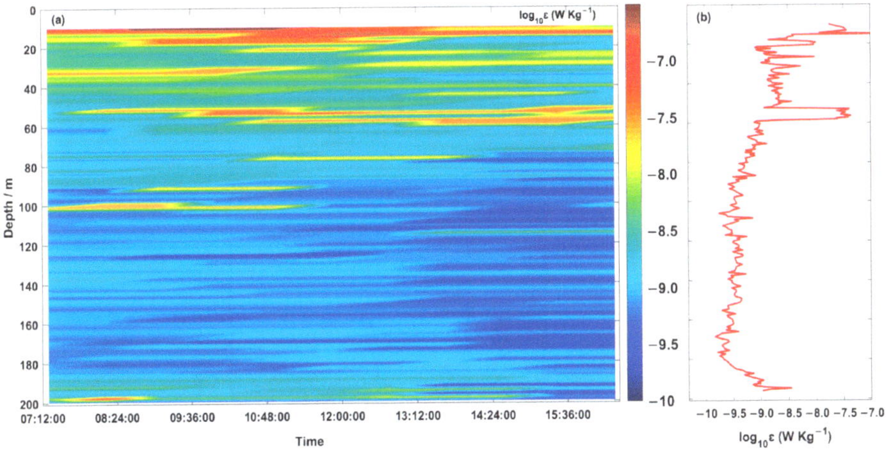


**Figure 9.** Depth–time maps of (**a**) the dissipation rate and (**b**) the temporally averaged dissipation rate.

The intensity of diapycnal mixing produced by observations of turbulent kinetic energy dissipation is estimated using the Osborn relation for each $\varepsilon$ estimate:

$$K_\rho = \Gamma\, \varepsilon / N^2, \tag{4}$$

where $\Gamma$ is a dimensionless mixing efficiency, $N = \sqrt{(g/\rho_0)\left(\frac{d\rho}{dz}\right)}$ is the buoyancy frequency (here $g$ and $\rho_0$ are the gravity and reference density, respectively, $\rho$ is the potential density profile, and $z$ is the depth). Here, we used the conventional constant of mixing efficiency, $\Gamma = 0.2$ [15,26].

In the depth–time map of the diapycnal diffusivity in Figure 10a, a thin surface mixed layer is visible, with an averaged $K_\rho$ of $10^{-5}$ $m^2{\cdot}s^{-1}$ and a strongly stratified thermocline with an averaged $K_\rho$ of $10^{-5.5}$ $m^2{\cdot}s^{-1}$ (approximately 40 m~150 m). Below the thermocline, $K_\rho$ displayed the same pattern as $\varepsilon$. The periods of relatively low and high $K_\rho$ were well-matched with the periods of low and high $\varepsilon$, respectively (compare Figures 9b and 10b), and the variations of $K_\rho$ were strongly correlated with the variations of $\varepsilon$.

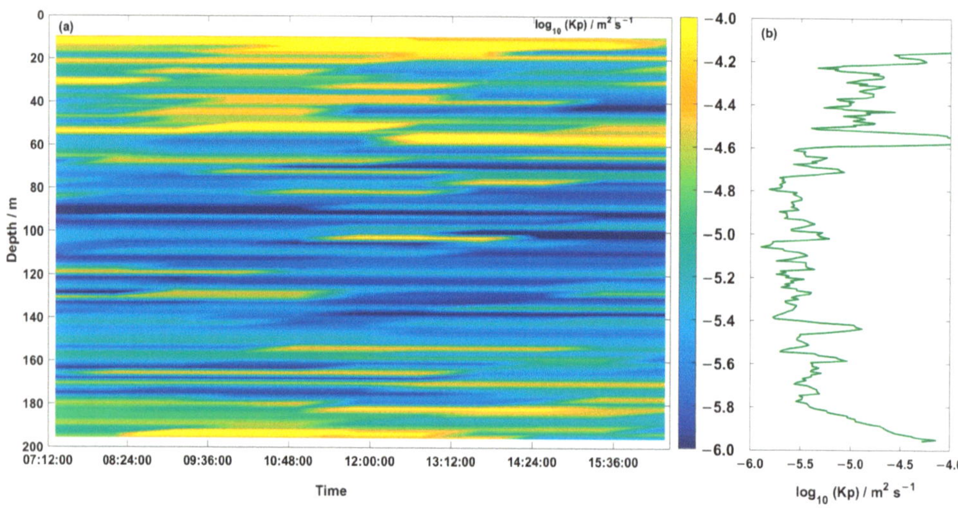


**Figure 10.** Depth–time maps of (**a**) diapycnal diffusivity and (**b**) the temporally averaged diapycnal diffusivity.

The dissipation rate ($\varepsilon$) in this section varied from $1.41 \times 10^{-10}$ to $4.18 \times 10^{-7}$ $W{\cdot}kg^{-1}$. The larger turbulent energy dissipation occurred in the surface mixing layer, which was larger than that of the thermocline and bottom mixing layer by about three orders of magnitude. Due to the effects of sea surface wind energy and buoyancy flux, the vertical mixing of the surface mixing layer was large and uniform. In the upper boundary of the thermocline, there was a violent mixing process, and the change of turbulent energy dissipation rate and mixing rate was very obvious. With the increase in depth, the mixing below the thermocline gradually weakened, in the order of $10^{-10}$~$10^{-9.5}$ $W{\cdot}kg^{-1}$. Therefore, the thermocline was like a "barrier." Above the "barrier" was an active open area, where wind energy and thermal radiation provided energy for strong turbulence mixing, whereas below the "barrier" turbulence mixing became very weak. The strong and stable thermocline in the water hindered the transfer of wind and heat energy from the surface mixed layer to the deep layer. The variation range of diapycnal diffusivity was $10^{-6}$~$10^{-4}$ $m^2{\cdot}s^{-1}$, and was higher between the surface mixing layer and the upper boundary of thermocline. These results are consistent with those of Shang et al. (2017) [27], who also found that the turbulent dissipation rate and diapycnal diffusivity would decrease under the thermocline "barrier."

## 6. Conclusions

In this study, a hybrid long-range multi-motion AUV was developed and used to measure turbulence measurement in the *nSCS*. The hybrid AUV was operated in a combined glider and cruise mode to sample turbulence with high resolution in both the vertical and horizontal directions. A total of ten continuous profiles were completed covering a horizontal span of 25 km and a depth of 200 m. The hybrid AUV's average speed ($U$) during descent and ascent was 0.57 $m \cdot s^{-1}$ and 0.65 $m \cdot s^{-1}$, respectively, while in cruise mode it was 0.68 m/s. The hybrid AUV's flight performance was stable and satisfied the requirement for turbulence observation. The orthogonally installed shear probes recorded the velocity shear with good agreement, and the noise-reduced shear spectra from both probes were in excellent agreement with the Nasmyth spectrum. The lowest detection level of the dissipation rate was $<3 \times 10^{-10}$ $W \cdot kg^{-1}$, which is comparable to that of the majority of vertical microstructure profilers.

The water column in the study area was highly stratified, with an average thick thermocline extending from 40 to 150 m below the surface. The dissipation rate ($\varepsilon$) in this section (Figure 9) varied from $1.41 \times 10^{-10}$ to $4.18 \times 10^{-7}$ $W \cdot kg^{-1}$. High values of $\varepsilon$ ($10^{-9} \sim 10^{-8}$ $W \cdot kg^{-1}$) were observed in the surface mixed layer, pointing toward the sea surface. In the upper boundary of the thermocline, there was a violent mixing process, and the change of dissipation rate and diapycnal diffusivity was very obvious (Figures 9b and 10b). In the thermocline, there was obvious variation in the dissipation rate and diapycnal diffusivity. The average dissipation rate and diapycnal diffusivity in the thermocline were $10^{-9.5} \sim 10^{-9}$ $W \cdot kg^{-1}$ and $10^{-5.5}$ $m^2 \cdot s^{-1}$, respectively, which were smaller than the level of the surface mixed layer. Therefore, the thermocline was like a "barrier," above which was an active open area, with wind energy and thermal radiation providing energy for strong turbulence mixing, and below which turbulence mixing became very weak. The strong and stable thermocline in the water damped the transfer of energy.

Overall, the measurement of turbulence from hybrid AUVs is a powerful addition to traditional turbulence instruments, as they make it possible to survey over long periods and large areas with high spatial resolution in both vertical and horizontal directions. Furthermore, the measurements are reasonably accurate and require much less dedicated time.

**Author Contributions:** Conceptualization, Y.N. and X.L. (Xin Luan); methodology, D.S.; software, Y.N.; validation, Y.H.; formal analysis, L.D.; investigation, Y.N. and Y.H.; writing—original draft preparation, Y.N.; writing—review and editing, X.L. (Xiuyan Liu); project administration, D.S. All authors have read and agreed to the published version of the manuscript.

**Funding:** This research was funded by the Fundamental Research Funds for the Central Universities, grant number 201962010, Shandong Province Double-Hundred Talent Plan, grant number WST2020002, the National Natural Science Foundation of China, grant number 62001262, and the Natural Science Foundation of Shandong Province, grant number ZR2020QF008.

**Institutional Review Board Statement:** Not applicable.

**Informed Consent Statement:** Not applicable.

**Data Availability Statement:** Not applicable.

**Conflicts of Interest:** The authors declare no conflict of interest.

## Nomenclature

| | |
|---|---|
| AUV | Autonomous underwater vehicle |
| *nSCS* | the northern South China Sea |
| CPMTM | Cross-platform instrument for microstructure turbulence measurements |
| CTD | Conductivity, temperature, depth |
| DVL | Doppler velocity log |
| R/V | Research vessel |

| | |
|---|---|
| AOA | Angle of attack |
| PSU | Practical salinity units |
| FFT | Fast-Fourier transform |
| $\varepsilon$ | The dissipation rate |
| $\theta$ | The pitch angle |
| $\gamma$ | The glide angle |
| $\alpha$ | The angle of attack |
| $U$ | The hybrid AUV speed through water |
| $P$ | Pressure |
| $W$ | The vertical speed |
| $k$ | The wavenumber |
| $K\rho$ | The intensity of diapycnal mixing |

## References


1. Holappa, K.W. *Ocean Turbulence Measurement Using an Autonomous Underwater Vehicle*; Florida Atlantic University: Boca Raton, FL, USA, 1997.
2. Creed, E.; Ross, W.; Lueck, R.; Stern, P.; Douglas, W.; Wolk, F.; Hall, R. Integration of a RSI microstructure sensing package into a Seaglider. In Proceedings of the OCEANS 2015—MTS/IEEE Washington 2015, Washington, DC, USA, 19–22 October 2015.
3. Lueck, R.G.; Wolk, F.; Yamazaki, H. Oceanic velocity microstructure measurements in the 20th century. *J. Oceanogr.* **2002**, *58*, 153–174. [CrossRef]
4. Grant, H.L.; Stewart, R.W.; Moilliet, A. Turbulence spectra from a tidal channel. *J. Fluid Mech.* **1962**, *12*, 241–263. [CrossRef]
5. Grant, H.L.; Moilliet, A.; Vogel, W.M. Some observations of the occurrence of turbulence in and above the thermocline. *J. Fluid Mech.* **1968**, *34*, 443–498. [CrossRef]
6. Dhanak, M.R.; Holappa, K. An autonomous ocean turbulence measurement platform. *J. Atmos. Ocean. Technol.* **1999**, *16*, 1506–1518. [CrossRef]
7. Greenan, B.J.W.; Oakey, N.S. A tethered free-fall glider to measure ocean turbulence. *J. Atmos. Ocean.* **1999**, *16*, 1545–1555. [CrossRef]
8. Goodman, L.; Wang, Z. Turbulence observations in the northern bight of monterey bay from a small AUV. *J. Mar. Syst.* **2009**, *77*, 441–458. [CrossRef]
9. Wolk, F.; Lueck, R.G.; Laurent, L.S. Turbulence measurements from a glider. In Proceedings of the OCEANS 2009, MTS/IEEE Biloxi—Marine Technology for Our Future: Global and Local Challenges, Biloxi, MS, USA, 26–29 October 2009.
10. Todd, R.E. High-frequency internal waves and thick bottom mixed layers observed by gliders in the Gulf Stream. *Geophys. Res. Lett.* **2017**, *44*, 6316–6325. [CrossRef]
11. Nagai, T.; Inoue, R.; Tandon, A.; Yamazaki, H. Evidence of enhanced double-diffusive convection below the main stream of the Kuroshio Extension. *J. Geophys. Res. Ocean.* **2015**, *120*, 8402–8421. [CrossRef]
12. Nie, Y.; Song, D.; Wang, Z.; Huang, Y.; Yang, H. Design and Motion Performance Analysis of Turbulent AUV Measuring Platform. *Sensors* **2022**, *22*, 460. [CrossRef] [PubMed]
13. Huang, Y.; Qiao, J.; Yu, J.; Wang, Z.; Xie, Z.; Liu, K. Sea-Whale 2000: A Long-range Hybrid Autonomous Underwater Vehicle for Ocean Observation. In Proceedings of the OCEANS 2019-Marseille, Marseille, France, 17–20 June 2019.
14. Nie, Y.; Yang, H.; Song, D.; Huang, Y.; Liu, X.; Hui, X. A new cross-platform instrument for microstructure turbulence measurements. *J. Mar. Sci. Eng.* **2021**, *9*, 1051. [CrossRef]
15. Osborn, T.R.; Crawford, W.R. An airfoil probe for measuring turbulent velocity fluctuations in water. In *Air-Sea Interaction;* Springer: Boston, MA, USA, 1980.
16. Goto, Y.; Yasuda, I.; Nagasawa, M. Turbulence Estimation Using Fast-Response Thermistors Attached to a Free-Fall Vertical Microstructure Profiler. *J. Atmos. Ocean. Technol.* **2016**, *33*, 2065–2078. [CrossRef]
17. Tian, J.; Yang, Q.; Zhao, W. Enhanced Diapycnal Mixing in the South China Sea. *J. Phys. Oceanogr.* **2010**, *39*, 3191. [CrossRef]
18. Fer, I.; Peterson, A.K.; Ullgren, J.E. Microstructure Measurements from an Underwater Glider in the Turbulent Faroe Bank Channel Overflow. *J. Atmos. Ocean. Technol.* **2014**, *31*, 1128–1150. [CrossRef]
19. Merckelbach, L.; Smeed, D.; Griffiths, G. Vertical water velocities from underwater gliders. *J. Atmos. Ocean. Technol.* **2010**, *27*, 547–563. [CrossRef]
20. Levitus, S. *Climatological Atlas of the World Ocean*; Noaa Professional Paper; Eos, Transactions American Geophysical Union: Washington, DC, USA, 1983; Volume 64, pp. 962–963.
21. Goodman, L.; Levine, E.R.; Lueck, R.G. On measuring the terms of the turbulent kinetic energy budget from an AUV. *J. Atmos. Ocean. Technol.* **2006**, *23*, 977–990. [CrossRef]
22. Nasmyth, P.W. Oceanic Turbulence. Ph.D. Thesis, Department of Physics, University of British Columbia, Vancouver, BC, Canada, 1970; p. 69.
23. Wolk, F.; Yamazaki, H.; Seuront, L.; Lueck, R.G. A new free-fall profiler for measuring biophysical microstructure. *J. Atmos. Ocean. Technol.* **2002**, *19*, 780. [CrossRef]
24. Gregg, M.C. Estimation and geography of diapycnal mixing in the stratified ocean. *Coast Estuar. Stud.* **1998**, *54*, 305–338.

25. Polzin, K.L.; Toole, J.M.; Schmitt, R.W. Finescale parameterizations of turbulent dissipation. *J. Phys. Oceanogr.* **1995**, *25*, 306–328. [CrossRef]
26. Thorpe, S. *An Introduction to Ocean Turbulence*; Cambridge University Press: Cambridge, UK, 2007; 264p.
27. Shang, X.D.; Liang, C.D.; Chen, G.Y. Spatial distribution of turbulent mixing in the upper ocean of the south china sea. *Ocean Sci.* **2017**, *13*, 503–519. [CrossRef]

sensors

MDPI

*Article*

# A Simple Approach to Estimate the Drag Coefficients of a Submerged Floater

**Yuval Hoffman, Liav Nagar, Ilan Shachar and Roee Diamant** *

Hatter Department of Marine Technologies, University of Haifa, Haifa 3498838, Israel
* Correspondence: roee.d@univ.haifa.ac.il

**Abstract:** The calculation of the drag force is a fundamental requirement in the design of any submerged system intended for marine exploration. The calculation can be performed by analytic analysis, numerical modeling, or by a direct calculation performed in a designated testing facility. However, for complex structures and especially those with a non-rigid design, the analytic and numerical analyses are not sufficiently accurate, while the direct calculation is a costly operation. In this paper, we propose a simple approach for how to calculate the drag coefficient in-situ. Aimed specifically at the complex case of elastic objects whose modeling via Computer-Aided Design (CAD) is challenging, our approach evaluates the relation between the object's speed at steady-state and its mass to extract the drag coefficient in any desired direction, the hydro-static force, and, when relevant, also the thruster's force. We demonstrate our approach for the special case of a highly complex elastic-shaped floater that profiles the water column. The analysis of two such floaters in two different sea environments shows accurate evaluation results and supports our claim for robustness. In particular, the simplicity of the approach makes it appealing for any arbitrary shaped object.

**Keywords:** drag force; hydrostatic force; thruster force; submerged floaters

## 1. Introduction

Floaters are a valuable tool for probing the water column. They can provide water current estimation [1] by tracking their drift motion over time, characterize internal waves [2] by observing spatial changes in the depth of the bathymetrical layer, or monitor changes in the marine environment by measuring temperature profiles over time with fine resolution [3]. Examples of floaters include profiling floats such as the Argo floats [4], which traverse the water column from the seabed to the sea surface for conductivity–temperature–depth (CTD) measurements, and Lagrangian floats, which are designed to drift with the water current [5]. For both types, calculating the drag coefficient is important information for improving the system design. Floaters are also used to evaluate the water current's properties [6]. Here, the analysis of the floater's drag is essential to filter out friction forces from the measurements. The drag is an attribute of how well the float drifts in the water current and is a function of the float's shape and size. A perfect, rounded, neutrally buoyant float with a narrow edge on its endfire will sense little resistance as it traverses through the water, while a square float with a large surface facing the drifting direction will sense high friction-like resistance.

The calculation of the drag coefficient of a submerged object is performed either by analytical modeling, using numerical finite-elements simulations, or directly in designated testbeds. The first analytical expression to calculate the drag coefficient is the seminal work by Jean le Rond d'Alembert in 1752, who calculated the drag force of a flow acting on a cylinder shape object using the potential flow theory (incompressible and inviscid, for a Reynolds number $Re >> 1$). The analysis yielded a zero drag force, which was in contrast to the experiments made. This mismatch is referred to as the *D'Alembert's paradox* and is due to neglecting the water boundary layers. The later theoretical work of Munk and

Glauert in the 1920s [7] presented the base for *thin airfoil theory*, which uses potential flow theory to calculate the drag and lift force on a thin wind. For Re << 1, for which the inertia at the N–S equation can be neglected, the so-called *stock flow* or the *creeping flow* linear equation is obtained, which can be solved in a close form to yield the drag coefficient [8]. That being said, analytics methods are limited to simple geometries and flow regimes, and current methodology leans towards numerical evaluation.

In the practical case where the object studied is of complex geometry or when the object travels through a complex flow regime, a common technique to determine the drag coefficients is the Computational-Fluid Dynamics (CFD) method [9]. The framework of CFD allows the estimation of the drag coefficients directly from its 3D model and thus allows performance evaluation even before its manufacturing to reduce the price of design. The method fragments the fluid volume into discrete cells, where each cell equation (i.e., Navier–Stokes, Euler equations) is solved iteratively while respecting the overall boundary condition. As such, a CFD simulation takes a long time to converge and requires advanced computing equipment. Moreover, for objects of non-rigid structures, such as a floater with many sharp edges and a flexible surface design, the model of the object may be too simplified, and the numerical calculation may lead to a significant error. In such cases, designers turn to direct drag measurements [10].

Measurements of drag coefficients are mostly performed in a tow tank (cf. [10]) to test the drag over the object itself or its smaller scale model. This kind of testing facility involves a Planar Motion Mechanism (PMM) that is stationed above the tank. This is an electromechanical device used for maneuvering studies, which moves in a specific regime while directly measuring the force induced on the object tested. While results are accurate, the process requires a large and expensive facility that is only profitable for shipyards or large companies. An alternative approach involves the use of a reduced-order model (ROM). Proposed by both Morrison and Younger [11] and Cely et al. [12], the hydrodynamic coefficient is calculated using a simple experiment. The tested object, in this case a remotely operated vehicle, is mounted on springs inside a water tank to measure its free decay. From the free decay measurements, the damping coefficients is extracted using non-linear least squared (NL-LS) methods. A similar approach by Eng et al. [13] uses a pendulum ROM to calculate the hydrodynamic coefficient. Still, the ROM offers only an approximation. To solve this, alternative approaches are offered to experimentally calculate the drag coefficient.

The use of experiments offers the possibility to test the object in realistic conditions. Caccia et al. [14] demonstrated the calculation of the drag of a thruster-aided vehicle in a sea experiment. The method involves measuring both the depth of the vehicle and its thruster's power. Then, by measuring the voltage–thrust relation in a complementary experiment, the method accounts for the propeller–hull and propeller–propeller effects that are usually neglected in standard numerical models. Still, a hard assumption is made claiming the drag force is omnidirectional, even though it may hold strong directional dependencies [15]. For example, this assumption may break when the object includes moving or flexible parts or has a non-uniform section area body. As such, the above approach is limited to a rigid body with a uniform section area.

In this work, we offer a low-complexity approach for the in-situ measurement of the drag coefficient. Our method defines a calibration test for a tested vehicle to measure the relation between its motion and its weight. Since the steady state velocity of the vehicle is a function of its drag, we evaluate the parameters of the speed–weight relation to extract the drag force and the hydrostatic force. To get a good estimation of the former relation, the calibration test should include traveling at different speeds. We solve this by performing multiple runs with different weights on the vehicle. We showcase our approach on a self-made floater that uses a thruster to ascend and a parachute to slow down its descending speed, thus decreasing the floater's energy consumption. Being long and narrow, the drifter's surface-to-weight ratio reduces the effect of the environment. To examine this, we tested the operation of the floater in two different sea experiments and analyzed its ascending and descending motion to yield a calculation of the drag. The results show a match between the relation assumed in our analysis between the squared speed and the floater's mass and a fair accuracy when comparing the estimated values for the hydrostatic force and thruster force to their direct measurement. Our novelty is in the concept of using weights to yield a different ascending/descending speed for the drifter without changing the other parameters of the system. In this way, we could harness the relation between the speed and drag. We argue that, without using weights, to evaluate the drag, one should control the system actively, through e.g., an active control mechanism, which would complicate the trial greatly up to the point that it could not be performed in-situ.

The structure of this paper is as follows. A literature review from Instrumentation and Measurement journals is offered in Section 2. In Section 3, we introduce our system model, main assumptions, and the structure of our floater as a test case. In Section 4, we describe our in-situ approach for measuring the drag coefficient. Results from two sea experiments are presented in Section 5, and conclusions are drawn in Section 6.

## 2. Survey of Instrumentation and Measurements Techniques

Considerations of how to measure the drag force over a tested system in a wind tunnel are presented in [16], where the dimensions of the measuring unit are emphasized. Aiming for a small testing unit, a suspension force-measuring system (SFMS) is proposed in [17]. Results show a sensitivity of 0.25 mV/N for drag force measurements, but a long calibration process is required. The measuring of the drag force is similar in essence to that of the friction force. Techniques for accurately measuring the friction coefficient are surveyed in [18,19]. For example, a common standard in this field is described in [20], where acoustic, optical, or tread sensors are installed in the runway to measure friction-related parameters. The existing methods require testing facilities, and gaps are identified regarding how to consider environmental conditions in the measurement. This is specifically important when operating in a sea environment.

The measurement of the drag coefficient for marine applications has been the focus of many instrumentation research works. In [21], the turbulence statistical method has been adopted to measure the drag force of a piezoelectric-based device for the wall friction drag reduction of micro underwater vehicles. The tests include a source for turbulence burst, a current meter, and the placement of the tested device downstream at a target Reynolds number. Particle image velocimetry is used to visualize the fluid and measure the turbulence. The impact of the drag can also be observed from the relation between the speed of the tested vehicle and the feedback received from its motor, namely the motor's current and the propeller's speed. This is similar to the method in [22] that maps between this relation based on fuzzy logic and an adaptive filter. Another form of impact of the drag is investigated in [23] to design a non-linear dynamic controller for AUVs. The review in [24] describes the different considerations in designing an underwater vehicle. For example, in [25], a design of a glider designed to glide smoothly in the water is proposed, and a controller that changes the vehicle's angle of attack to reduce the drag, for example, is proposed. Similarly, design concepts based on drag for AUVs are described in [26].

A few techniques exist for measuring the drag and its impact on underwater vehicles. In [27], the effect of the drag is measured in a pool experiment in a lab for an underwater fan-wing thruster for both towing and self-driving modes. The vehicle velocity and the vertical source are analyzed for the latter, whereas a special device for measuring the drag force is used for the former. Results are found to be in agreement with simulation analysis. In [28], experiments measure the hydrodynamic forces on an AUV as a function of its acceleration and velocity. Static measurement in a straight line and dynamic tests involving pitch changes are compared to measurements in a water tunnel. However, the method requires stability and thus does not apply in realistic sea conditions except for deep sea tests. To prove the design of an AUV with soft legs and a soft inflatable morphing body for position keeping, ref. [29] measured the drag and lift of a vehicle in a flow channel using an actuator and a servo motor as well as laser Doppler velocimetry for velocity measurement.

Operating at sea, a horizontal drag measurement platform was used in [30] for turbulence observation with the help of an extremely stable platform. A method to evaluate the active and passive drag of a swimmer is proposed in [31]. The method relies on residual thrust measurements, i.e., the difference between the propulsive and resistive forces, performed in a water flume that allows the flow velocity to be adjusted along with an evaluation of the relation between the swimmer's propulsion and drag. In [32], the Levenberg–Marquardt algorithm is used to estimate the drag coefficient of an AUV in a calibrated pitot tube. While accurate measurements are obtained, to the best of our knowledge, there is no robust method for estimating the drag of any object in-situ in realistic sea conditions.

## 3. System Model

Our system includes a complex-shaped floater that profiles the water column using a thruster while collecting acoustic and CTD measurements. The floater includes means to measure its instantaneous depth and is time-synchronized prior to deployment. The calibration of the floater's drag involves a few ascend/descend cycles before changing the floater's weights. Then, the floater's depth profile is analyzed offline.

As illustrated in Figure 1, the floater considered is a 3-inch cylinder containing electronic components: a hydrophone; pressure, salinity, and temperature sensors; and a thruster. The floater's depth profiling is performed by defining a lower and upper limit to the floater. Specifically, being negatively buoyant, the floater operates its thruster to ascend, and stops the thruster to descend. Then, upon reaching a lower depth limit, the floater operates its thruster and ascends, and upon reaching its upper depth limit, the floater stops the thruster and descends. To reduce battery usage, the floater includes a parachute-like tarpaulin sheet that, much like an umbrella, opens when the floater sinks and closes when it ascends. This is made possible by three rigid elements given the fabric flexibility against the water pressure. As such, the floater holds its complex shape, and its drag coefficient is hard to calculate numerically. In particular, its flexibility makes the floater's boundary conditions complex to evaluate. A picture of the floater during operation is shown in Figure 2.

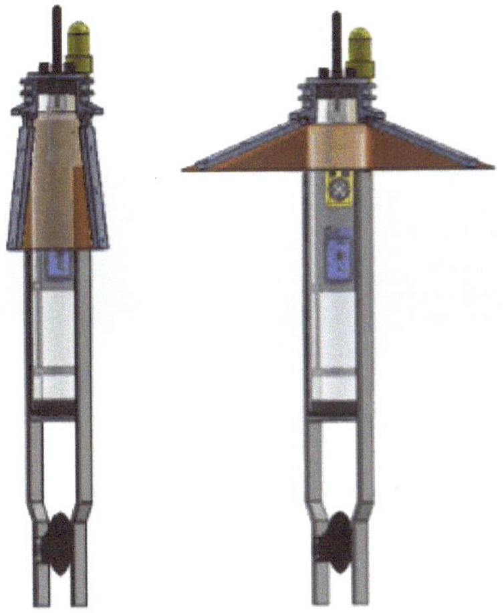

**Figure 1.** A model of the considered floater. In the left panel, the floater's arms are close, while in the right one, they are open.

**Figure 2.** A picture of the floater in operation with a scuba diver in the background for scale. In this picture, the parachute-like tarpaulin sheet is in a semi-open stage.

## 4. Our Drag Coefficient Measuring Technique

The key idea behind our method is the utilization of the relation between the floater's steady-state speed and its weight. We perform the drag estimate in steady state to eliminate the drag coefficient dependency in the system's acceleration. Since there is a square relation between the drag force and the vehicle's velocity, the ratio between the vehicle's speed and its weight should be also a square relation. Then, by quantifying the parameters of the latter relation, we can evaluate the drag force as well as the floater's hydrostatic force. The specific test thus includes allowing the floater to rise and submerge while collecting depth profiles to evaluate the floater's speed and repeating this process multiple times with different weights to yield several points for an offline speed–weight ratio evaluation.

### *4.1. Method Formalization*

Let $\rho_p$ be the plumbum density, $v_p$ be the volume of the floater's weights, $\rho_w$ be the water density, and $v_b$ be the volume of the floater. Also let, $C_{up}$ be the quadratic drag coefficient, $A_{\text{up}}$ be the floater's section area while rising, and $F_{\text{thruster}}$ be the force obtained from the thruster. The force balance on the floater while rising is

$$\begin{aligned} F_{\text{x}} &= (m+\rho_{\text{p}}v_{\text{p}})\ddot{x} = -mg+\rho_{\text{w}}v_{\text{b}}g\breve{}(\rho_{\text{p}}-\rho_{\text{w}})v_{\text{p}}g \\ &- \frac{1}{2}\rho_{\text{w}}C_{\text{up}}A_{\text{up}}\dot{x}^2 + F_{\text{thruster}}\,, \end{aligned} \tag{1}$$

where $m$ is the floater's mass.

#### 4.1.1. Steady State: Ascending

To formalize the steady-state behavior, the following definitions are required. The quadratic damping coefficient is

$$\Psi_{\text{up}} = \frac{1}{2}\rho_{\text{w}}C_{\text{up}}A_{\text{up}}\,. \tag{2}$$

The total hydrostatic force on the floater body is

$$\Lambda = -mg+\rho_{\text{w}}v_{\text{b}}g\,. \tag{3}$$

The total hydrostatic force acting on the sinkers is

$$\kappa = (\rho_{\text{p}}-\rho_{\text{w}})v_{\text{p}}g\,. \tag{4}$$

Then, we rewrite (1) as

$$(m+\rho_{\text{p}}v_{\text{p}})\ddot{x} = \Lambda-\kappa-\Psi_{\text{up}}\dot{x}^2+F_{\text{thruster}}\,. \tag{5}$$

At steady state, the velocity is

$$\dot{x}^2 = -\frac{\kappa}{\Psi_{\text{up}}}+\frac{\Lambda+F_{\text{thruster}}}{\Psi_{\text{up}}}\,. \tag{6}$$

Note that in our calibration experiments, the right-most term at the right hand side of (6) is constant, while the other terms are variable.

#### 4.1.2. Steady-State: Descending

When the floater submerges, the drag force changes its direction, and the floater's force, $F_{\text{thruster}}$, vanishes. Then,

$$\begin{aligned} F_x &= (m+\rho_{\text{p}}v_{\text{p}})\ddot{x} = -mg+\rho_{\text{w}}v_{\text{b}}g \\ &- (\rho_{\text{p}}-\rho_{\text{w}})v_{\text{p}}g+\frac{1}{2}\rho_{\text{w}}C_{\text{down}}A_{\text{down}}\dot{x}^2\,, \end{aligned} \tag{7}$$

where $C_{\text{down}}$ and $A_{\text{down}}$ are the quadratic drag coefficient and the floater's section area while descending, respectively. The damping coefficient can then be written as

$$\Psi_{\text{down}} = \frac{1}{2}\rho_{\text{w}}C_{\text{down}}A_{\text{down}}\,. \tag{8}$$

We rewrite (7) as

$$(m+\rho_{\text{p}}v_{\text{p}})\ddot{x} = \Lambda-\kappa+\Psi_{\text{down}}\dot{x}^2\,, \tag{9}$$

and the descending steady state becomes

$$\dot{x}^2 = \frac{\kappa}{\Psi_{\text{down}}}-\frac{\Lambda}{\Psi_{\text{down}}}\,. \tag{10}$$

The above analysis shows a linear relation between $\dot{x}^2$ and $\kappa$. This linear relation holds with small Reynolds numbers, as in the case of the drifter. We use this observation as a validation metric in the following.

## 5. Experimental Evaluation

### 5.1. Experimental Setup

To demonstrate the applicability of our approach in realistic scenarios, we performed two sea experiments using two different floaters. The first experiment was conducted in the Mediterranean Sea in September 2021 across the shores of Haifa at a shallow water environment of depth 10 m. The sea state was 2, and the water current was roughly 2 knots. The floater was deployed from a small vessel and performed ascend/descend profiles at the depth range of 1 m and 7 m. After five profiles, the floater automatically ascended to the surface where additional plumbum weights were added. We note that, due to their dense mass with respect to the magnitude of the water current, the drag of the weights can be neglected. The process then repeated itself for two more runs of five profiles each. In Table 1, we give the mass of the added weights and the measured in-water mass for this experiment. During the entire operation, the floater was observed from the distance by a snorkeler who made sure that, during its profiles, the floater did not touch the seabed or reach the surface. An example of the depth profile as collected by the floater is shown in Figure 3.

**Table 1.** Weights' mass and underwater mass during the Haifa experiment.

| Profile Number | Weights' Mass [g] | Underwater Weight [N] |
|---|---|---|
| 1 | 129 | 1.15 |
| 2 | 363 | 3.23 |
| 3 | 816 | 7.3 |

The second experiment took place in the Red Sea, Eilat, Israel in January 2022. This experiment was conducted in deeper water at 30 m depth. A different floater than the one used in the first experiment but with a similar shape was handled by scuba divers who descended to 20 m depth and operated the floater. The floater continuously performed profiles between 5 m depth and 20 m, and every five ascend–descend profiles, the divers changed the floater's weights. The use of different weights between the two experiments was to compensate for the different salinities of the two seas explored, as well as to add diversity to the results. We note that the drifter reached a steady state speed after less than 1 m from the maximum point of rising or descending, and thus testing at lower depth had no influence on performance except for additional data points for better speed estimation via regression. In total, five weight changes were made, and the experiment lasted for 50 min. The mass of the added weights and the measured in-water mass for this experiment are given in Table 2. An example of the depth profile as collected by the floater is shown in Figure 4.

**Table 2.** Weights' mass and underwater mass during the Eilat experiment.

| Profile Number | Weights' Mass [g] | Underwater Weight [N] |
|---|---|---|
| 1 | 210 | 1.877 |
| 2 | 380 | 3.396 |
| 3 | 550 | 4.915 |
| 4 | 720 | 6.435 |
| 5 | 890 | 7.954 |

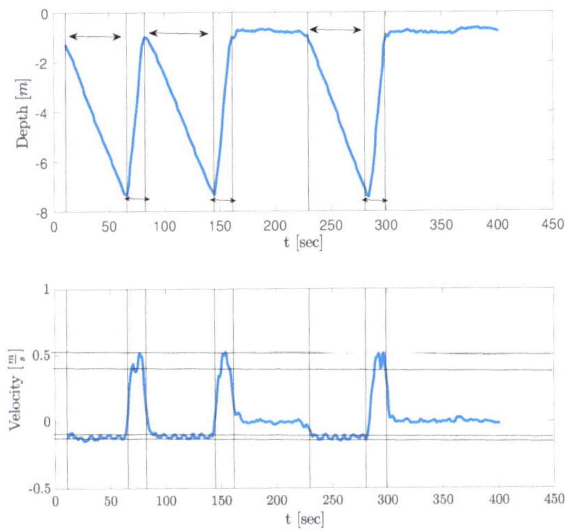

**Figure 3.** Example of three depth profiles recorded by the floater during the Haifa sea experiment. Arrows mark rising and descending periods.

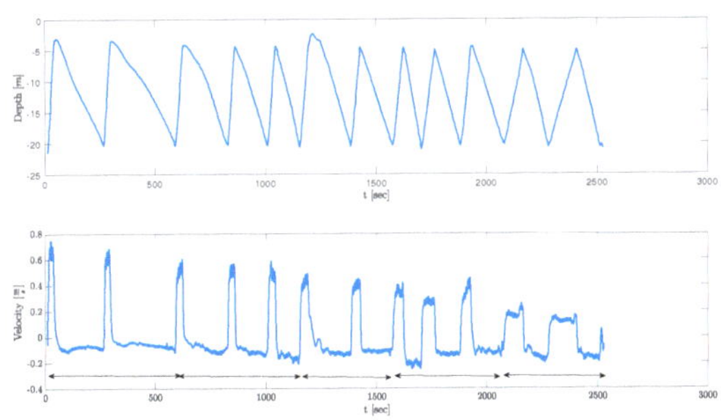

**Figure 4.** Example of multiple depth profiles recorded by the floater during the Eilat sea experiment. Arrows mark periods for each set of weights.

The outcome of the two experiments was a measure of the relation between the floaters' measured velocity at steady state, $\dot{x}$, and the measured mass of their plumbum weights, $\kappa$. To that end, the floaters' initial buoyancy, i.e., before adding weights, was made negative, causing the floater to sink with no thruster force—that is, $\Lambda - \kappa < 0$. The offline analysis included averaging the five ascend–descend profiles to yield the average steady-state ascend and descend speed.

### 5.2. Experimental Results

By (6) and (10), the relation between the velocity and the weights' mass is quadratic. To parametrize this relation and to extract the drag force, we performed a linear regression over the measured square velocity and the weights' mass. The results of this linear fitting for the Haifa trial, along with the averaged measurements, are shown in Figure 5a. We observe a good fit between the measured relation and the linear fitting from which the following parameters are drawn: $\Psi_{\text{up}} = 23.9$ kg/m, $\Psi_{\text{down}} = 188.5$ kg/m, $\Lambda = -0.68$ N, $F_{\text{thruster}} = 7.07$ N. The results for the Eilat trial are given in Figure 5b. These results also show a good fit for the linearization attempt, and the following parameters are drawn: $\Psi_{\text{up}} = 16.8$ kg/m, $\Psi_{\text{down}} = 301.2$ kg/m, $\Lambda = -1.11$ N, $F_{\text{thruster}} = 7.4$ N.

We note the large differences in the estimated drag force for the descending and ascending directions caused by the operation of the parachute. Comparing the results between the two experiments, we observe a change in the measured drag force (ascending and descending). This difference is due to a different parachute size (for descending) and a different thruster (for ascending).

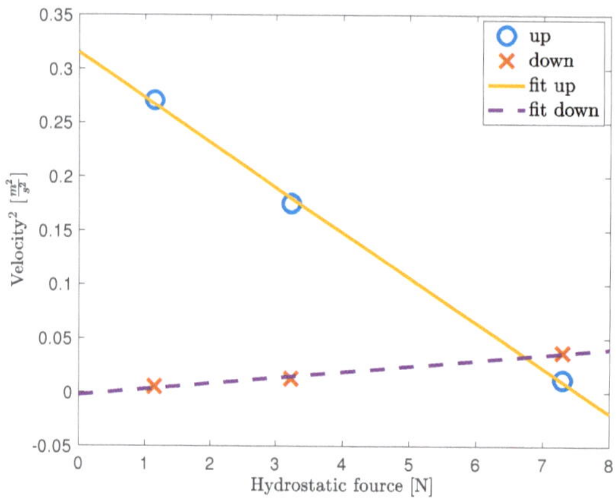


(**a**)

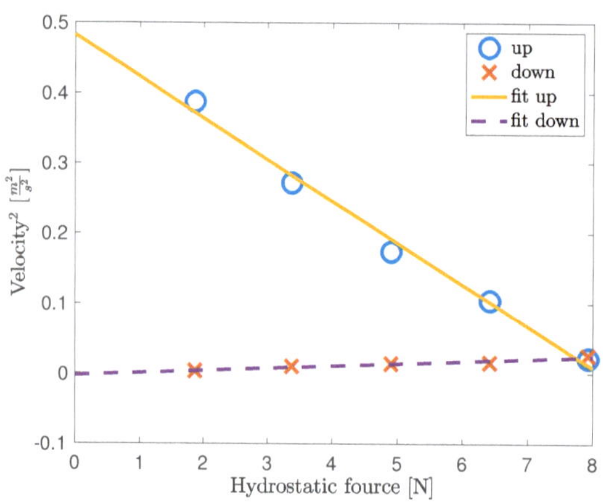


(**b**)

**Figure 5.** Squared floater's velocity vs. weights' mass: measured and linear fit. Results are shown separately for ascending and descending. (**a**) Haifa, (**b**) Eilat.

### 5.3. Validation of Results

The small linearization error we observe in both sea experiments confirms our above analysis. Still, there is a need to validate the obtained results. Unfortunately, as noted in Section 1, directly measuring the drag coefficient of the two floaters poses a significant challenge since the flexibility of the parachute requires a steady-state measurement in both ascending and descending directions, which is hard to achieve in a testing facility. However, it is relatively straightforward to obtain a direct measurement of the hydrostatic force and the floater's thruster's force. Thus, we validate our results by comparing the direct and indirect measurement of these two latter characteristics.

The hydrostatic force can be measured by observing which is the minimum force required to make the floater sink. We provide a bound for this force by adding small weights to the floater as listed in Tables 1 and 2 until the floater sinks. In particular, we deployed the two floaters in a 9 m × 3 m salt water pool of 3 m depth and balanced the hydrostatic force to be slightly positive. After fine tuning, we observed that, when adding a weight of 129 grams to the floater used in the first experiment and a weight of 210 gram to the floater used in the second experiment, the hydrostatic force of the two floaters becomes negative and the floaters sink. As a result, the hydrostatic force can be bounded within the range $-1.15\ \text{N} \leq \Lambda < 0$ for the floater used in the first experiment and within $-1.877\ \text{N} \leq \Lambda < 0$ for the floater used in the second experiment. We note that a value in-between these ranges was obtained for both floaters in the in-situ measurements.

We note that a similar operation for measuring the thruster's force is harder since this value depends on the floater's speed. Still, assuming the speed difference along the floater's depth profile is negligible, we can compare the estimated thruster's force, $\Gamma_{\text{thruster}}$, to its static measurement. To that end, we balanced the floater at a slight positive hydrostatic force and measured the above-water height of the floater. We then operated the thruster—an operation that pushed the floater up and increased its above-water height. Comparing the above-water heights at the two operation modes, the thruster's force is evaluated by

$$\hat{F}_{\text{thruster}} = \Delta h \pi \frac{D^4}{4} \rho_{\text{w}} g\,, \tag{11}$$

where $\Delta h$ is the height difference between the thruster-on and the thruster-off modes, and $D$ is the floater's diameter. For the floater used in the first experiment, we obtained $\Delta h = 0.1$ m, which translates to $\hat{F}_{\text{thruster}} = 8.6$ N (vs. 7.07 N in the in-situ measurement). For the floater used in the second experiment, we obtained $\Delta h = 0.13$ m, which translates to $\hat{F}_{\text{thruster}} = 11.2$ N (vs. 7.4 N in the in-situ measurement). While in both cases, we argue that the results are very close, it is apparent that the analysis for the Haifa floater achieved better accuracy. This is because the floater used in the second experiment was equipped with a faster, and thus lower efficiency, thruster, which somewhat goes against our above assumption that the dynamic force and the static one are similar.

### *5.4. Discussion*

In this work, we have demonstrated an in-situ approach for measuring the drag coefficient in two sea environments and for two different floaters. The good results obtained in both cases support our claim for robustness and show that our approach can fit different depth profiles and can be implemented over different platforms. The difference between the values obtained for the two floaters, despite their seemingly similar shapes, emphasizes the need for an in-situ measurement such as ours rather than settling for numerical evaluation, especially when analyzing elastic shapes.

We admit that our method is limited by a few factors. First, the measurement of the object's speed should be performed in steady state, which requires careful planning of the calibration test. Second, the method is sensitive to strong water currents or waves that may change the structure of the examined object. For example, a strong water current may partially fold our floaters' parachute. Last, our assumption that the thruster's force in a dynamic scenario is similar to that in a static scenario only holds when the thruster operates at slow speed. As a result, the test for the drag coefficient must be planned as a calibration test rather than performed as part of an ongoing operation. To handle such discrepancies, future work may account for the system efficiency in the calculation of the speed–mass relation and for cases where the drag coefficient is not constant for different speeds.

## 6. Conclusions

In this work, we outline the details of an in-situ approach for calculating the drag force of an arbitrary shaped object. Our approach utilizes the expected quadratic relation between the object's speed and its mass. With no modeling of the object tested, and using a simple setup to evaluate the parameters of this relation, our method can accurately measure the object's drag force in any direction, its hydrostatic force, and the thruster's force if a thruster is included. Results from two sea experiments using thruster-operated floaters with an elastic shape demonstrated the applicability of our approach and revealed a good match between our in-situ evaluation and a direct measurement of the hydrostatic force and the thruster's force. Our solution is suitable to systems with a small Reynolds number and at steady state. Future work would explore how to extract the drag coefficient also in the non-linear regime.

**Author Contributions:** Conceptualization, Y.H. and R.D.; methodology, Y.H.; data analysis, Y.H. and R.D.; software, I.S.; hardware preparation L.N.; experimentation Y.H., I.S. and R.D.; data curation, R.D.; writing, Y.H. and R.D.; project administration, R.D.; funding acquisition, R.D. All authors have read and agreed to the published version of the manuscript.

**Funding:** This work was supported in part by the MOST–BMBF German–Israeli Cooperation in Marine Sciences 2018-2020 (Grant # 3-16573) and by the MOST action for Agriculture, Environment, and Water for the year 2019 (Grant # 3-16728).

**Institutional Review Board Statement:** Not applicable.

**Informed Consent Statement:** Not applicable.

**Data Availability Statement:** Not applicable.

**Conflicts of Interest:** The authors declare no conflict of interest. The funders had no role in the design of the study; in the collection, analyses, or interpretation of data; in the writing of the manuscript, or in the decision to publish the results.

## References

1. Santoki, M.; Ratheesh, S.; Sharma, R.; Joshipura, K.; Basu, S. Assimilation of drifter data in a circulation model of the Indian Ocean. *IEEE Geosci. Remote Sens. Lett.* **2011**, *9*, 100–103. [CrossRef]
2. Cao, J.; Lu, D.; Li, D.; Zeng, Z.; Yao, B.; Lian, L. Smartfloat: A multimodal underwater vehicle combining float and glider capabilities. *IEEE Access* **2019**, *7*, 77825–77838. [CrossRef]
3. Zhou, H.; Cao, J.; Fu, J.; Liu, C.; Wei, Z.; Yu, C.; Zeng, Z.; Yao, B.; Lian, L. Dynamic modeling and motion control of a novel conceptual multimodal underwater vehicle for autonomous sampling. *Ocean. Eng.* **2021**, *240*, 109917. [CrossRef]
4. Jayne, S.R.; Roemmich, D.; Zilberman, N.; Riser, S.C.; Johnson, K.S.; Johnson, G.C.; Piotrowicz, S.R. The Argo program: Present and future. *Oceanography* **2017**, *30*, 18–28. [CrossRef]
5. Van Sebille, E.; Griffies, S.M.; Abernathey, R.; Adams, T.P.; Berloff, P.; Biastoch, A.; Blanke, B.; Chassignet, E.P.; Cheng, Y.; Cotter, C.J.; et al. Lagrangian ocean analysis: Fundamentals and practices. *Ocean. Model.* **2018**, *121*, 49–75. [CrossRef]
6. Diamant, R. Prediction of water current using a swarm of submerged drifters. *IEEE Sensors J.* **2020**, *20*, 11598–11607. [CrossRef]
7. Abbott, I.H.; Von Doenhoff, A.E. *Theory of Wing Sections: Including a Summary of Airfoil Data*; Courier Corporation: New York, NY, USA, 2012.
8. Rong, L.; Dong, K.; Yu, A. A general solution of the drag force in packed beds of multi-sized particles under creeping fluid flow conditions. *Powder Technol.* **2017**, *315*, 87–97. [CrossRef]
9. Eça, L.; Hoekstra, M. A procedure for the estimation of the numerical uncertainty of CFD calculations based on grid refinement studies. *J. Comput. Phys.* **2014**, *262*, 104–130. [CrossRef]
10. Jagadeesh, P.; Murali, K.; Idichandy, V. Experimental investigation of hydrodynamic force coefficients over AUV hull form. *Ocean. Eng.* **2009**, *36*, 113–118. [CrossRef]
11. Morrison, A.; Yoerger, D.R. Determination of the hydrodynamic parameters of an underwater vehicle during small scale, nonuniform, 1-dimensional translation. In Proceedings of the OCEANS'93, IEEE, Victoria, BC, Canada, 18–21 October 1993; pp. II277–II282.
12. Cely, J.S.; Saltaren, R.; Portilla, G.; Yakrangi, O.; Rodriguez-Barroso, A. Experimental and computational methodology for the determination of hydrodynamic coefficients based on free decay test: Application to conception and control of underwater robots. *Sensors* **2019**, *19*, 3631. [CrossRef]
13. Eng, Y.; Lau, M.; Low, E.; Seet, G.; Chin, C. Estimation of the Hydrodynamics Coefficients of an ROV using Free Decay Pendulum Motion. *Eng. Lett.* **2008**, *16*, 326–331.
14. Caccia, M.; Indiveri, G.; Veruggio, G. Modeling and identification of open-frame variable configuration unmanned underwater vehicles. *IEEE J. Ocean. Eng.* **2000**, *25*, 227–240. [CrossRef]
15. Cheeseman, I. Fluid-Dynamic Drag: Practical Information on Aerodynamic Drag and Hydrodynamic Resistance. *Aeronaut. J.* **1976**, *80*, 371. [CrossRef]
16. Li, S.; You, Z.; Gao, H.; Wang, Q.; Wu, G.; Ma, G. Force Measurement and Support Integrated Device in Hypersonic Wind Tunnel. *IEEE Trans. Instrum. Meas.* **2022**, *71*, 9501709. [CrossRef]

17. Li, S.; Liu, Z.; Gao, H.; Lv, J.; Zhang, X. A New Airframe/Propulsion-Integrated Aerodynamic Testing Technology in Hypersonic Wind Tunnel. *IEEE Trans. Instrum. Meas.* **2022**, *71*, 7502809. [CrossRef]
18. Niu, Y.; Jiang, X.; Meng, F.; Wang, R.; Ju, G.; Zhang, S.; Meng, Z. Techniques and Methods for Runway Friction Measurement: A Review of State of the Art. *IEEE Trans. Instrum. Meas.* **2021**, *70*, 9510717. [CrossRef]
19. Panda, J.P.; Mitra, A.; Warrior, H.V. A review on the hydrodynamic characteristics of autonomous underwater vehicles. *Proc. Inst. Mech. Eng. Part J. Eng. Marit. Environ.* **2021**, *235*, 15–29. [CrossRef]
20. Batuwangala, E.; Silva, J.; Wild, G. The regulatory framework for safety management systems in airworthiness organisations. *Aerospace* **2018**, *5*, 117. [CrossRef]
21. Zhang, L.; Shan, X.; Yan, S.; Xie, T. Arrangement and control parameter selection methods for achieving high-efficiency underwater active drag reduction by micro piezoelectric actuator. *IEEE Trans. Ind. Electron.* **2022**, *70*, 1729–1739. [CrossRef]
22. Gao, S.; He, B.; Zhang, X.; Wan, J.; Mu, X.; Yan, T. Cruise Speed Estimation Strategy Based on Multiple Fuzzy Logic and Extended State Observer for Low-Cost AUV. *IEEE Trans. Instrum. Meas.* **2021**, *70*, 2502513. [CrossRef]
23. Tabataba'i-Nasab, F.S.; Keymasi Khalaji, A.; Moosavian, S.A.A. Adaptive nonlinear control of an autonomous underwater vehicle. *Trans. Inst. Meas. Control.* **2019**, *41*, 3121–3131. [CrossRef]
24. He, Y.; Wang, D.B.; Ali, Z.A. A review of different designs and control models of remotely operated underwater vehicle. *Meas. Control.* **2020**, *53*, 1561–1570. [CrossRef]
25. Xiaowei, L.; Yu, M.; Ping, O. Design of a kind of underwater glider. In Proceedings of the 2015 12th IEEE International Conference on Electronic Measurement & Instruments (ICEMI), Qingdao, China, 16–18 July 2015; Volume 2, pp. 688–692. [CrossRef]
26. Sánchez, P.J.B.; Papaelias, M.; Márquez, F.P.G. Autonomous underwater vehicles: Instrumentation and measurements. *IEEE Instrum. Meas. Mag.* **2020**, *23*, 105–114. [CrossRef]
27. Gao, T.; Lin, Y.; Ren, H.; Tse, Z.T.H. Hydrodynamic analyses of an underwater fan-wing thruster in self-driving and towing experiments. *Measurement* **2020**, *165*, 108132. [CrossRef]
28. Nouri, N.; Mostafapour, K.; Bahadori, R. An apparatus to estimate the hydrodynamic coefficients of autonomous underwater vehicles using water tunnel testing. *Rev. Sci. Instrum.* **2016**, *87*, 065106. [CrossRef]
29. Ishida, M.; Drotman, D.; Shih, B.; Hermes, M.; Luhar, M.; Tolley, M.T. Morphing Structure for Changing Hydrodynamic Characteristics of a Soft Underwater Walking Robot. *IEEE Robot. Autom. Lett.* **2019**, *4*, 4163–4169. [CrossRef]
30. Lueck, R.G.; Wolk, F.; Yamazaki, H. Oceanic velocity microstructure measurements in the 20th century. *J. Oceanogr.* **2002**, *58*, 153–174. [CrossRef]
31. Narita, K.; Nakashima, M.; Takagi, H. Developing a methodology for estimating the drag in front-crawl swimming at various velocities. *J. Biomech.* **2017**, *54*, 123–128. [CrossRef]
32. Jahangardy, Y.; Madoliat, R.; Nouri, N. Identification of drag force of the underwater vehicles. *J. Appl. Fluid Mech.* **2017**, *10*, 275–281. [CrossRef]

sensors

MDPI

*Article*

# USV-Observed Turbulent Heat Flux Induced by Late Spring Cold Dry Air Incursion over Sub-Mesoscale Warm Regions off Sanriku, Japan

**Akira Nagano [1,*], Takuya Hasegawa [2,3], Keisuke Ariyoshi [4], Takeshi Iinuma [4], Tatsuya Fukuda [5], Nobuhiro Fujii [6], Fumiaki Tomita [4,7] and Ryota Hino [8]**

1 Research Institute for Global Change, Japan Agency for Marine-Earth Science and Technology (JAMSTEC), Yokosuka 237-0061, Japan
2 Faculty of Environmental Earth Science, Hokkaido University, Sapporo 060-0808, Japan
3 Graduate School of Science, Tohoku University, Sendai 980-8577, Japan
4 Research Institute for Marine Geodynamics, Japan Agency for Marine-Earth Science and Technology (JAMSTEC), Yokohama 236-0001, Japan
5 Institute for Marine-Earth Exploration and Engineering, Japan Agency for Marine-Earth Science and Technology (JAMSTEC), Yokosuka 237-0061, Japan
6 Marine Works Japan Ltd., Yokosuka 237-0063, Japan
7 International Research Institute of Disaster Science, Tohoku University, Sendai 980-8572, Japan
8 Research Center for Prediction of Earthquake and Volcanic Eruptions, Graduate School of Science, Tohoku University, Sendai 980-8578, Japan
* Correspondence: nagano@jamstec.go.jp

**Abstract:** We performed oceanic and atmospheric observations in the region off the Sanriku coast, Japan, from May 11 to 5 July 2022, using a wave-propelled unmanned surface vehicle, a Wave Glider (WG). Despite the severe weather conditions of atmospheric low-pressure system crossings, we successfully measured wind, air temperature, humidity, and sea surface temperature over the course of 55 days to calculate the turbulent heat flux. The WG observed that the atmosphere became more humid due to the southerly wind along the northwestern rim of the North Pacific subtropical high. The warm Kuroshio water expanded to the southeast of Hokkaido as a result of the northward shedding of an anticyclonic mesoscale (~100 km) eddy, called a warm-core ring, from the Kuroshio Extension. The WG traversed smaller (sub-mesoscale) water regions that were warmer and saltier than the surrounding Kuroshio water. The observations indicate that cold, dry air masses advected by northerly winds following the passage of atmospheric low-pressure systems generate a substantial upward turbulent heat flux over sub-mesoscale warm water regions, contrasting to no heat flux in the surrounding Kuroshio water region.

**Keywords:** warm-core ring; air–sea interaction; turbulent heat flux; sub-mesoscale variation; Wave Glider

## 1. Introduction

A vast amount of heat is transported northward from the tropics to the subtropics via the Kuroshio, the western boundary current of the North Pacific subtropical gyre e.g., [1–3], which originates in the region east of the Philippines, e.g., [4] (Figure 1a). Thus, the sea surface temperature (SST) in the Kuroshio and Kuroshio Extension (KE) regions is extremely high compared to the surrounding regions, and heat is actively released into the atmosphere throughout the boreal winter as turbulent latent and sensible heat fluxes, e.g., [5,6]. The huge amount of heat released in the winter contributes to the global heat balance, e.g., [1]. Therefore, turbulent heat fluxes are key parameters for considering the role of the ocean in the climate system. In other seasons, a smaller but significant amount of heat is considered to be released to the atmosphere if the atmosphere and ocean are in favorable conditions. The released heat may have effects on Japan's regional climate and the weather. Knowledge on the turbulent heat flux variation leads to the improvement of climate and weather forecasts.

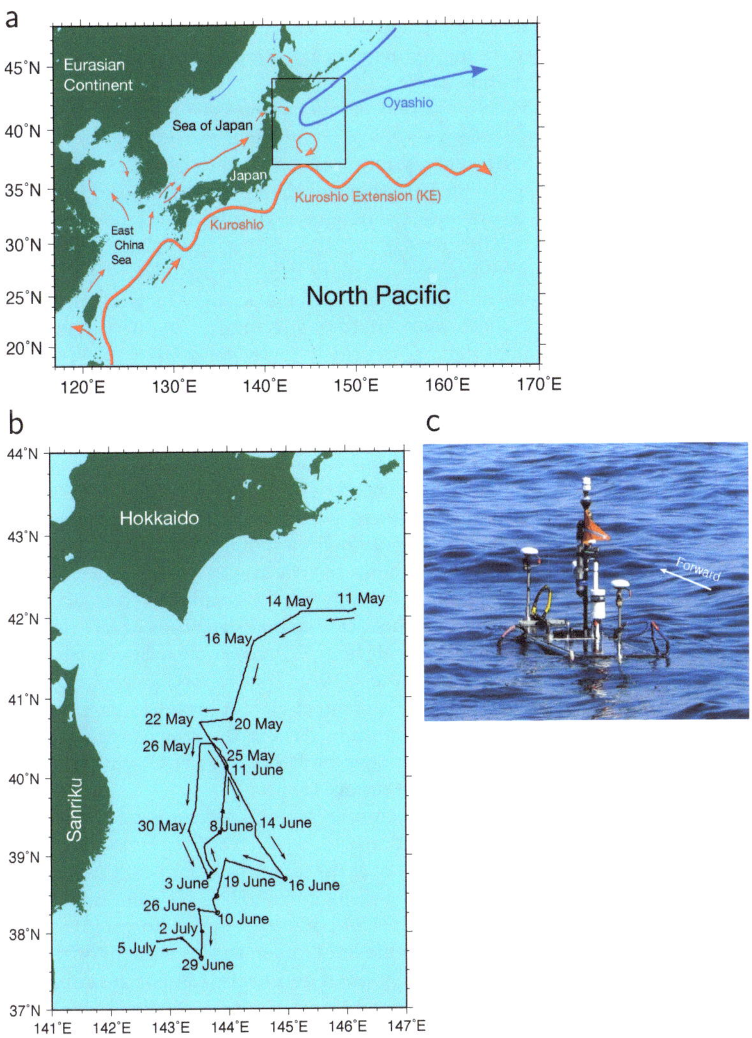


**Figure 1.** (**a**) Schematic diagram of the subtropical (red arrow lines) and subpolar (blue arrow lines) surface currents in the western North Pacific. The experiment region enclosed by the square is depicted by the enlarged map in panel (**b**). (**b**) Wave Glider (WG) track (black line) and passage dates. The directions of the WG are illustrated by arrows. (**c**) The observing WG immediately after the deployment on 11 May 2022.

The Oyashio, the western boundary current of the North Pacific subpolar gyre, flows southward in the region off the Hokkaido coast, Japan, (Figure 1a) and frequently proceeds to the east of the Sanriku coast. The Oyashio current transports fresh cold water to the eastern regions of Japan. The area between the Kuroshio and Oyashio regions is referred to as the perturbed area due to the complicated distribution of waters from the Kuroshio and Oyashio regions by mesoscale (30–100 km) eddies and fronts [7]. Strong mesoscale anticyclonic eddies, known as warm-core rings (WCRs), are occasionally detached from the KE meander's crest and intrude into the perturbed region (Figure 1a). The WCRs retain the Kuroshio water, characterized by high temperature and salinity, within their interiors, thereby exhibiting temperature and salinity contrasts in the region off the Sanriku coast. Furthermore, there may be smaller scale (sub-mesoscale) variations derived from the WCRs. The SST contrasts caused by mesoscale and sub-mesoscale variations have effects on the surrounding environment by generating circulations in the atmospheric boundary layer through the heat released into the atmosphere.

Lately, ocean mesoscale and sub-mesoscale variations have been mainly observed by satellites. By the state-of-the-art techniques, satellite observations provide high-resolution turbulent heat flux data with a horizontal resolution of 0.25°, which can resolve heat flux variations due to mesoscale eddies in the Kuroshio and KE regions [8]. However, certain variables, such as SST observed by satellites, are frequently obscured by clouds and other factors in rainy seasons even when the measurement resolutions are sufficiently high in space and time. In particular, sub-mesoscale variations in all the parameters required to estimate turbulent heat fluxes have not been observed in the perturbed region. Therefore, it is unclear how the oceanic sub-mesoscale variations affect the atmospheric boundary layer. In situ observations are important to accurately understand mesoscale and sub-mesoscale eddies, fronts, and their effects on the atmospheric boundary layer. However, massive research vessels could disrupt the boundary layers of the atmosphere and ocean. Smaller platforms are preferable to obtain accurate measurements of the variables required to estimate the turbulent latent heat flux (LHF) and sensible heat flux (SHF).

Unmanned surface vehicles (USVs) have been developed, e.g., [9], and are useful for observing atmospheric and oceanic parameters with high spatiotemporal resolutions and accuracy sufficient to support scientific research. For example, Nagano and Ando [6] observed the atmosphere and ocean across a mesoscale warm spot in the Kuroshio south of Japan using a Saildrone, a wind-powered USV (https://www.saildrone.com, accessed on 9 December 2022). They discovered smaller scale spatial structures of SST and the corresponding wind variation over the mesoscale high SST region. By using ocean-wave-propelled USVs, called Wave Gliders (WGs) (https://www.liquid-robotics.com, accessed on 9 December 2022), and a research vessel, Nagano et al. [10] observed variations in atmospheric and oceanic variables on time scales of days in the eastern region of the western tropical Pacific warm pool. Observations by USVs can reveal sub-mesoscale variations in SST and other variables at the sea surface layer.

Wind speed, relative humidity, air temperature, and SST are required to estimate turbulent heat flux based on a bulk flux algorithm, e.g., [11,12]. Of these atmospheric and oceanic parameters, measurements of humidity over the sea are especially difficult because sensors are not designed for use at sea and are easily damaged by seawater immersion. Nevertheless, humidity is a particularly important parameter in the Kuroshio and KE regions throughout the year, associated with the Baiu (Japan's rainy season) onset and withdrawal in late spring to early summer, typhoons and heavy rainfall in summer to autumn, and cold dry winds from the Eurasian Continent in winter to early spring.

To observe the variation in the turbulent heat flux on spatial scales down to the sub-mesoscale in the perturbed region, we performed a field experiment by deploying a WG in the region off the Hokkaido coast on 11 May 2022, observing for 55 days, and recovering it on 5 July 2022. To obtain reliable humidity data throughout the observation period, we protected humidity and air temperature sensors by covering them with pieces of waterproof breathable fabric, as described below. We measured all the parameters required to estimate turbulent heat flux from the bulk flux algorithm during severe weather conditions caused by the passage of several atmospheric low-pressure systems over the WG, and as a result, we obtained turbulent heat fluxes east of the Sanriku coast. In particular, we successfully observed sub-mesoscale warm waters in the perturbed area for the first time. Additionally, we observed an increase in specific humidity during the observation period both before and during Japan's rainy season, i.e., the Baiu season, accompanied by moistened southerly winds and substantial heat release from the ocean to the atmosphere caused by occasional occurrences of dry cold northerly winds over the sub-mesoscale warm regions.

The remaining sections of this study are structured as follows. The observation and configuration of the WG are described in Section 2. In Section 3, we calculate turbulent heat flux based on the parameters observed by the WG and, using satellite and reanalysis data, identify the parameters and conditions where the heat flux is enhanced. The summary and conclusions are provided in Section 4.

## 2. Observation and Data

In addition to the primary object of the WG observation, which was the observation of crustal deformation using the global navigation satellite system-acoustic (GNSS-A) technique as reported by Iinuma et al. [13], we installed meteorological and oceanographical sensors on the WG and measured atmospheric and oceanic variables. The USV used in this experiment is a Wave Glider (version SV3) (Liquid Robotics, Sunnyvale, CA, USA; https://www.liquid-robotics.com, accessed on 9 December 2022), which is a form of autonomous vehicle powered by waves [14,15]. The underwater glider, which is connected to the float by an 8-m umbilical cable, converts wave energy into forward motion. The float has a length and width of 3.05 and 0.81 m, respectively. Details of the WG used in this experiment are provided by Iinuma et al. [13]. The WG was deployed in the region southeast of Hokkaido on 11 May 2022 at 07:40 (Japanese standard time [JST]) by the research vessel (R/V) *Kaiyo Maru No. 3* (Kaiyo Engineering Co., Ltd., Tokyo, Japan) and recovered on 5 July 2022 at 05:40 (JST) by the R/V *Kaiyo Maru No. 2* (Kaiyo Engineering Co., Ltd., Tokyo, Japan). During the observation, the GNSS determined the location of the WG with a 2-m uncertainty radius. The WG's path is depicted in Figure 1b. The WG observed meteorological and oceanographic data along the path, occasionally staying at GNSS-A stations to conduct the observations of crustal deformation.

A conductivity-temperature-depth (CTD) JES10mini (Offshore Technologies, Yokohama, Japan) was installed at a depth of 0.2 m under the WG's float (Figure 2) to collect temperature and salinity data at 1-min intervals. The temperature and salinity accuracies of the CTD are $\pm 0.005$ °C and $\pm 0.005\ \mathrm{S\,m^{-1}}$, respectively. The accuracy of salinity is better than the $\pm 0.04$ (practical salinity scale). The salinity readings declined commencing around 15 June 2022, to a minimum of 31.5, despite being in the salt-rich Kuroshio region, and then increased to over $\sim$33 on 26 June 2022. This abnormally low salinity may have been caused by a goose barnacle stuck on the inside of the glass tube containing the electrodes, which inhibited the intake of seawater. Accordingly, the salinity data after 15 June 2022 were omitted from the subsequent analysis.

A 200WX Weather Station (AIRMAR Technology Corporation, Milford, NH, USA) was equipped to the middle mast of the WG at a height of 1.1 m (Figure 2), and air temperature, barometric pressure, wind direction, and speed were observed at 10-min intervals. Additionally, we installed three HOBO U23-001A Pro v2 data loggers (Onset Computer Corporation, Bourme, MA, USA) to the head (0.5 m height), middle (0.9 m height), and tail (0.5 m height) masts, and we observed air temperature and relative humidity at every 5-min interval. The accuracies of temperature and relative humidity were $\pm 0.2$ °C and $\pm 2.5\%$, respectively. The information of these meteorological and oceanographic instruments is listed in Table 1.

We doubly covered the HOBO sensors with commercially available Gore-Tex (W.L. Gore & Associates, Inc., Flagstaff, AZ, USA) fabric to prevent seawater from damaging the humidity sensors. The Gore-Tex fabric used in this experiment features a three-layered structure with a waterproof, breathable membrane of expanded polytetrafluoroethylene sandwiched between chemical synthetic fabrics. An examination by use of a diode laser spectroscope shows that Gore-Tex fabric is quite breathable for water vapor [16]. Although the WG was once overturned after deployment, relative humidity sensors continued to function properly throughout the observation.

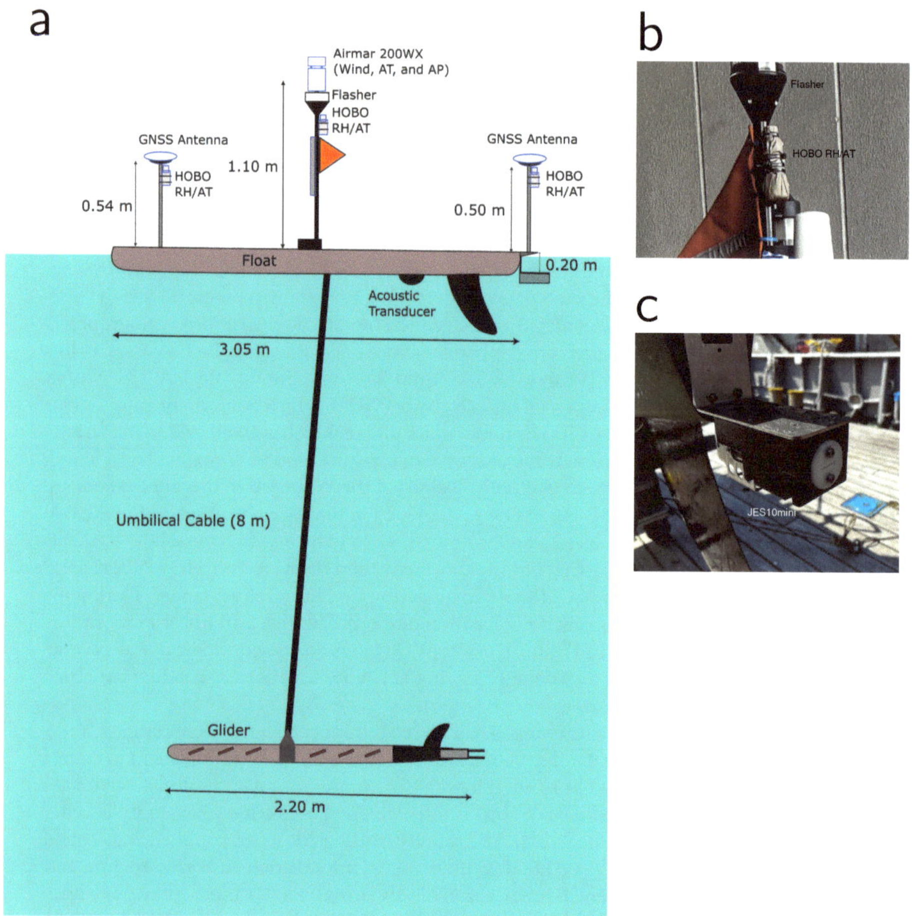


**Figure 2.** (**a**) Diagram of a Wave Glider SV3. Images of (**b**) the HOBO data logger on the middle mast of the WG and (**c**) the CTD JES10mini under the tail of the WG float. The HOBO sensor in panel (**b**) is covered by beige Gore-Tex fabric.

Temperatures inside and outside the fabric are assumed to change by $\Delta T_\mathrm{i}$ and $\Delta T_\mathrm{o}$, respectively, while air moves through the Gore-Tex fabric at an exchange rate of $E$ for a duration of $\Delta t$. Considering the heat balance inside the fabric of volume $V$, we obtain the ratio of $\Delta T_\mathrm{i}$ to $\Delta T_\mathrm{o}$, i.e., $\Delta T_\mathrm{i}/\Delta T_\mathrm{o} = 1/(1 + V/(E\Delta t))$. In the absence of fabric, we expect $E$ to be almost infinite, implying that the ratio would be infinitely close to the unity. We compared the temperatures collected by the Weather Station and the HOBO sensors. We observed a correlation coefficient of $r = 0.94$ (Figure 3), which is significantly higher than the 99% confidence level, and the root-mean-square error was 1.55 °C. The regression slope (0.77), which is equivalent to $\Delta T_\mathrm{i}/\Delta T_\mathrm{o}$, is less than one. Thus, the exchange of air may be reduced by a finite $E$ due to the existence of the Gore-Tex fabric. The same argument may apply to the moisture concentration, but we did not measure the relative humidity outside the fabric. If our guess is correct, the humidity variation observed within the fabric may be approximately 33% lower than expected. Nevertheless, the HOBO humidity measurements are useful for monitoring variations in humidity.

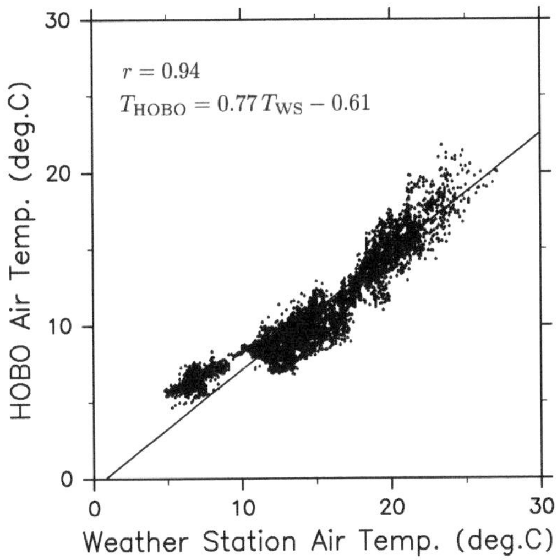


**Figure 3.** Scatter plot of observed temperature values by Weather Station sensor ($T_{WS}$) versus those by HOBO sensor ($T_{HOBO}$). Their correlation coefficient is $r = 0.94$. The linear regression equation ($T_{HOBO} = 0.77\,T_{WS} - 0.61$) is indicated by the slanted line.

We computed LHF and SHF by applying the COARE3.0b bulk flux algorithm developed by Fairall et al. [12] to atmospheric and oceanic variables collected every 10 min by the WG. We used temperature data from the JES10mini CTD sensor as SST data to calculate heat fluxes rather than performing the skin temperature correction, as the shortwave and longwave radiation fluxes were not measured.

Daily $1/4^\circ \times 1/4^\circ$ gridded SST data from May to July 2022 (NOAA OI SST V2 High Resolution Dataset) provided by Reynolds et al. [17] were used to monitor horizontal SST distribution in the study region spanning 37–44° N, 141–147° E. We employed the sea-level pressure field in the region of 30–45° N, 130–160° E obtained from $1.25^\circ \times 1.25^\circ$ gridded data from the Japanese 55-year Reanalysis (JRA-55) conducted by the Japan Meteorological Agency [18].

**Table 1.** Information of meteorological and oceanographic instruments installed to the Wave Glider.

| Measurement | Model | Manufacturer | Accuracy | Resolution |
|---|---|---|---|---|
| Wind Speed<br>Wind Direction<br>Air Pressure<br>Air Temperature | 200WX<br>Weather<br>Station | AIRMAR<br>Technology | ±5%<br>±3°<br>±0.5 hPa<br>±1.1 °C | 0.1 m s$^{-1}$<br>0.1°<br>0.1 hPa<br>0.1 °C |
| Air Temperature<br>Relative Humidity | HOBO U23<br>Pro v2<br>Data Logger | Onset<br>Computer | ±0.2 °C<br>±2.5% | 0.04 °C<br>0.05% |
| Water Temperature<br>Conductivity<br>Water Pressure | JES10mini | Offshore<br>Technologies | ±0.005 °C<br>±0.005 S m$^{-1}$<br>±0.1% FSR | 0.0001 °C<br>0.00001 S m$^{-1}$ |

## 3. Results and Discussion

The SST observed by the WG increased with time throughout the observation period (Figure 4a). Particularly, on 18 May 2022, when the WG crossed the SST front of the Oyashio south of Hokkaido, the SST increased significantly. An SST differential of more than approximately 5 °C can be observed across the front. Intriguingly, air temperature (Figure 4b) and specific humidity (Figure 4c) substantially increased along with the increase in SST, when the WG crossed the Oyashio front. Thus, the atmospheric boundary layer is significantly affected by the horizontal distribution of SST around the Oyashio.

Except for the period from mid-May to mid-June, the WG detected south to south-westerly winds (Figure 4d), consistent with the climatological wind direction in this season. The southerly winds along the northwestern rim of the North Pacific subtropical high transport moistened air mass from the south, e.g., [19], resulting in an increase in specific humidity (Figure 4c). During the period from mid-May to mid-June, atmospheric low-pressure systems encircled Japan and altered the wind speed and direction east of the Sanriku coast (Figure 4d). The passage of the low-pressure systems caused the atmospheric boundary layer to become cold and dry, as indicated by the low air temperature and low specific humidity in Figure 4b,c. The relationship between the low-pressure systems and turbulent heat flux will be discussed in detail below.

In Figure 5, the LHF and SHF are represented by red and blue lines, respectively. Substantial heat transfer from the ocean to the atmosphere occurred during the yellow-banded periods of decreasing air temperature and specific humidity. Meanwhile, slight heat absorption by the ocean (cyan band in Figure 5) was observed during the period when the WG was stationed north of the Oyashio front (cyan band in Figure 4a) in the low SST region.

Maps of satellite-observed SST for individual heat absorption and release events marked, respectively, by cyan (a) and yellow (b–e) bands in Figure 5 are depicted in Figure 6. In early to mid-May, 2022, the coastal Oyashio water colder than 10 °C extended to just east of the Sanriku coast (Figure 6a). The warm Kuroshio water extended east of the Sanriku coast (Figure 6b,c) and reached the southern coast of Hokkaido (Figure 6d) after shedding an anticyclonic mesoscale eddy. i.e., a WCR, located approximately 39.4° N, 144.5° E from the KE (Figure 6a). In other words, as discovered by Hasegawa et al. [20], the cold coastal Oyashio current retreated to the east of Hokkaido, accompanied by the northward shedding of the WCR.

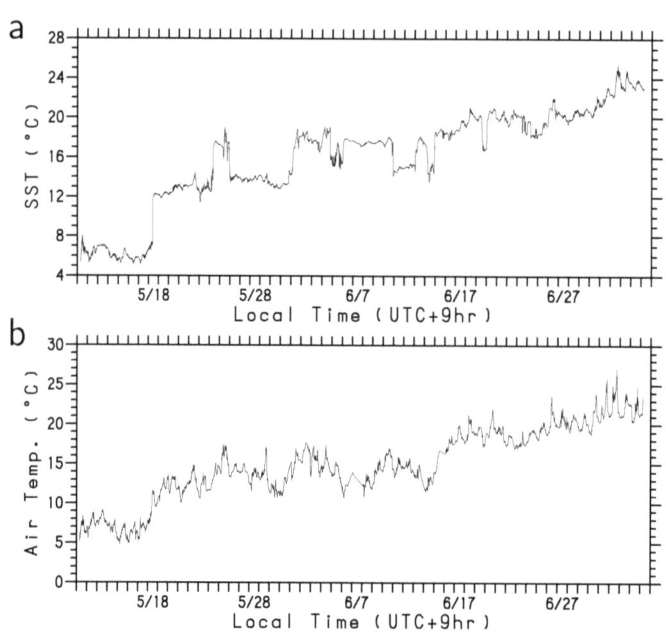


**Figure 4.** *Cont.*

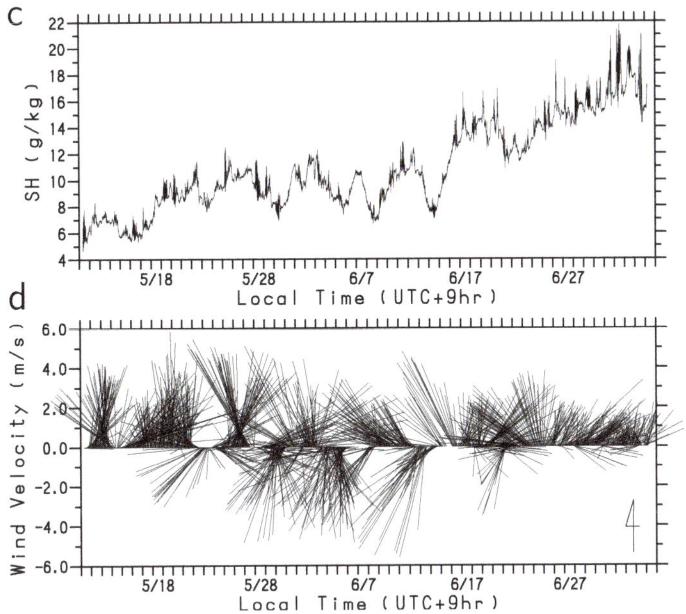


**Figure 4.** Time series of (**a**) SST (°C), (**b**) air temperature (°C), (**c**) specific humidity ($g\,kg^{-1}$), and (**d**) wind velocity vector ($m\,s^{-1}$). The abscissa is local time (UTC + 9 h) during the observation period from 11 May to 5 July 2022.

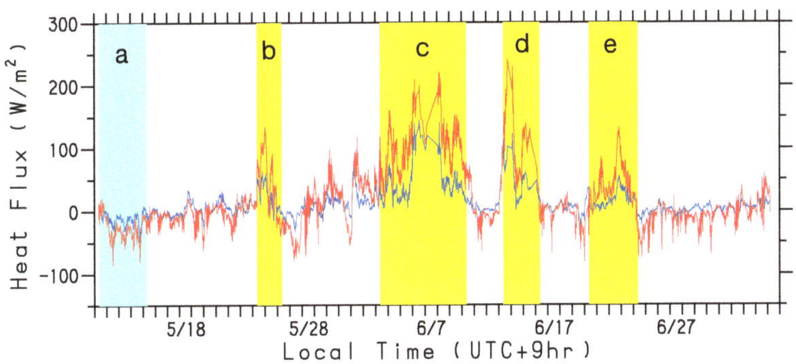


**Figure 5.** Time series of latent (red line) and sensible (blue line) sea surface heat fluxes ($W\,m^{-2}$) along the WG track. The abscissa is local time (UTC + 9 h) during the observation period from 11 May to 5 July 2022. Cyan and yellow bands marked by letters a–e indicate periods during which heat fluxes are shown in Figure 6.

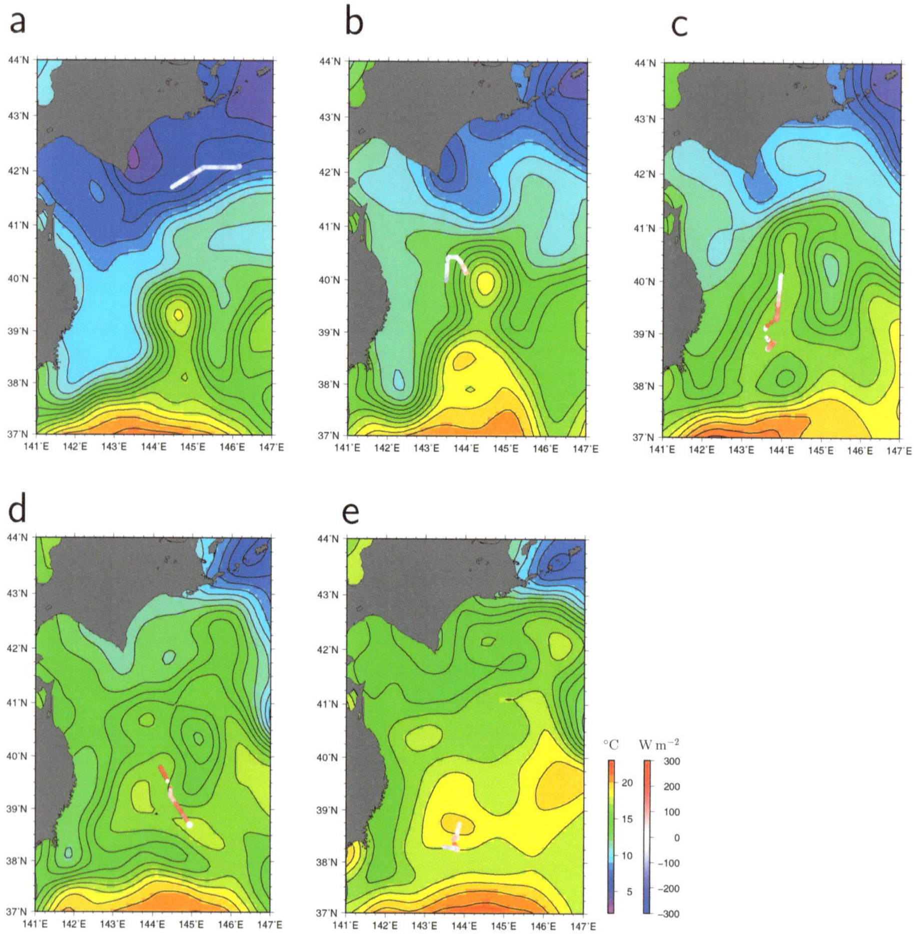


**Figure 6.** The Wave Glider tracks during the period shown by letters a–e in Figure 5, which correspond to panel letters of this figure. Values of turbulent (latent plus sensible) heat flux (W m$^{-2}$) are shown by colors of the WG tracks. Maps of SST (°C) are satellite-observed values on (**a**) 13 May, (**b**) 25 May, (**c**) 7 June, (**d**) 15 June, and (**e**) 22 June 2022. The contour intervals of SST are 1 °C.

The gradual increase in the observed SST by the WG after 1 June 2022 is partly attributable to the northward extension of the Kuroshio water. Although the WG had been in the warm Kuroshio water since 18 May, heat from the ocean was not always transferred to the atmosphere; instead, it was only deprived after when atmospheric disturbances passed over the WG (Figure 5). Furthermore, we identified high SST signals that are likely caused due to sub-mesoscale warm water regions in the Kuroshio water, such as those that occurred from 24 to 25 May, 31 May to 10 June, 13 to 14 June, and 21 to 23 June (Figure 4a). These signals correspond to high sea surface salinity (SSS) signals with the exception of 21 to 23 June (Figure 7). They cannot be fully detected by satellite observations because of the low resolution (Figure 6). The upward turbulent heat flux was notably large (>100 W m$^{-2}$ and >50 W m$^{-2}$ for LHF and SHF, respectively) in most of these sub-mesoscale high SST (>18 °C) regions (marked by letters b, c, d, and e in Figure 5). Although the WG was in very warm (∼20 °C) water during 21 to 23 June (Figure 6e), the heat flux (letter e in Figure 5) was lower than in early and mid-June (letters c and d) because the air temperature and specific humidity were relatively high. This WG observation in the perturbed region revealed that sub-mesoscale warm waters provide significant heat to the atmosphere in late spring.

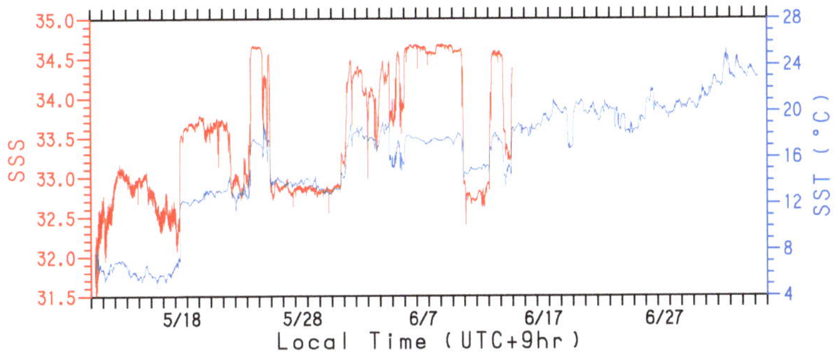


**Figure 7.** Time series of SSS (red line) in practical salinity scale along the WG track. The abscissa is local time (UTC + 9 h) during the observation period from 11 May to 5 July 2022. For comparison, SST (°C) time series is shown by blue line.

In accordance with Nagano et al. [6,10], we decompose the heat flux variation $F$ into components due to SST ($T_S$), air temperature ($T_A$), specific humidity ($q$), and wind speed ($U$), in order to examine the variables that are responsible for the variations in LHF and SHF, as follows:

$$\Delta F \approx \left(\frac{\partial F}{\partial T_S}\right)\Delta T_S + \left(\frac{\partial F}{\partial T_A}\right)\Delta T_A + \left(\frac{\partial F}{\partial q}\right)\Delta q + \left(\frac{\partial F}{\partial U}\right)\Delta U, \quad (1)$$

where $\Delta T_S$, $\Delta T_A$, $\Delta q$, and $\Delta U$ are temporal variations in SST, air temperature, specific humidity, and wind speed, respectively. Due to the complexity of the turbulent heat flux function, the rate of change of each term on the right-hand side of Equation (1) was determined using the mean values of the other three variables during the observation period. Units of all the terms of Equation (1) are the same as that of turbulent heat flux, i.e., $\mathrm{W\,m^{-2}}$.

The variation components in LHF and SHF due to the four variables are depicted in Figure 8a,b, respectively. However, the wind speed variation was not primarily responsible for the increase in turbulent heat flux (orange lines in Figure 8). This is because the strong winds were accompanied by a relatively high air temperature (Figure 4b) and high specific humidity (Figure 4c). Throughout the observation period, the effects of the increase in SST (magenta line in Figure 8) on LHF and SHF were canceled out by the increases in specific humidity (Figure 8a) and air temperature (Figure 8b). The enhancements of the LHF (red line in Figure 5) and SHF (blue line) due to the atmospheric disturbances such as from 24 to 25 May, 1 to 10 June, and 13 to 15 June are attributed to the positive effects of specific humidity ($\sim$50–100 $\mathrm{W\,m^{-2}}$) (cyan line in Figure 8a) and air temperature ($\sim$10–50 $\mathrm{W\,m^{-2}}$) (green line in Figure 8b) in addition to the positive effect of sub-mesoscale high SST regions (magenta lines in Figure 8). The effect of sub-mesoscale high SST waters was variable up to $\sim$50 $\mathrm{W\,m^{-2}}$ ($\sim$30 $\mathrm{W\,m^{-2}}$) for the LHF (SHF) variation. Thus, although this humidity effect may be underestimated as described in Section 2, the humidity decreases are found to be particularly important for the enhancement of LHF in late spring.

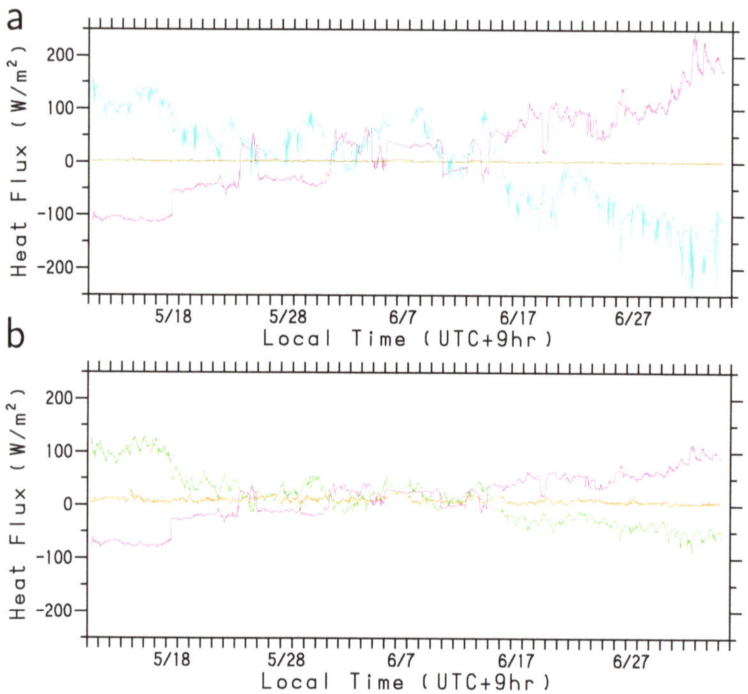


**Figure 8.** (**a**) LHF and (**b**) SHF variation components ($\mathrm{W\,m^{-2}}$) in Equation (1). The components of SST, air temperature, specific humidity, and wind speed are shown by magenta, green, cyan, and orange lines, respectively. Because SHF is independent of specific humidity, the SHF variation due to specific humidity is not plotted in panel (**b**). The abscissa is local time (UTC + 9 h) during the observation period from 11 May to 5 July 2022.

Figure 9 illustrates sea-level pressure maps during the peak periods of the turbulent heat flux. The turbulent heat flux was positively large (marked by letters by b, c, and d in Figure 5) except on 23 June 2022, when the atmospheric low-pressure systems proceeded to the east of the WG and wind blew southward (Figure 9a–c). Although the winds were strong during the passage of these low-pressure systems, high air temperature and specific humidity prevented heat from being released into the atmosphere. Prior to the rise of the turbulent heat flux (letter c in Figure 5), the SSS was low and highly variable from 1 to 4 June (Figure 7), occasionally reaching below 33. The low-SSS signal corresponds to the passage of the low pressure and may be the result of heavy rainfall. This is consistent with the high specific humidity. Due to the high atmospheric pressure to the northeast of Hokkaido on June 23 (Figure 9d), relatively dry and low-temperature winds blew from the east and southeast as observed by the WG (Figure 4b,c), releasing heat into the atmosphere (letter e in Figure 5).

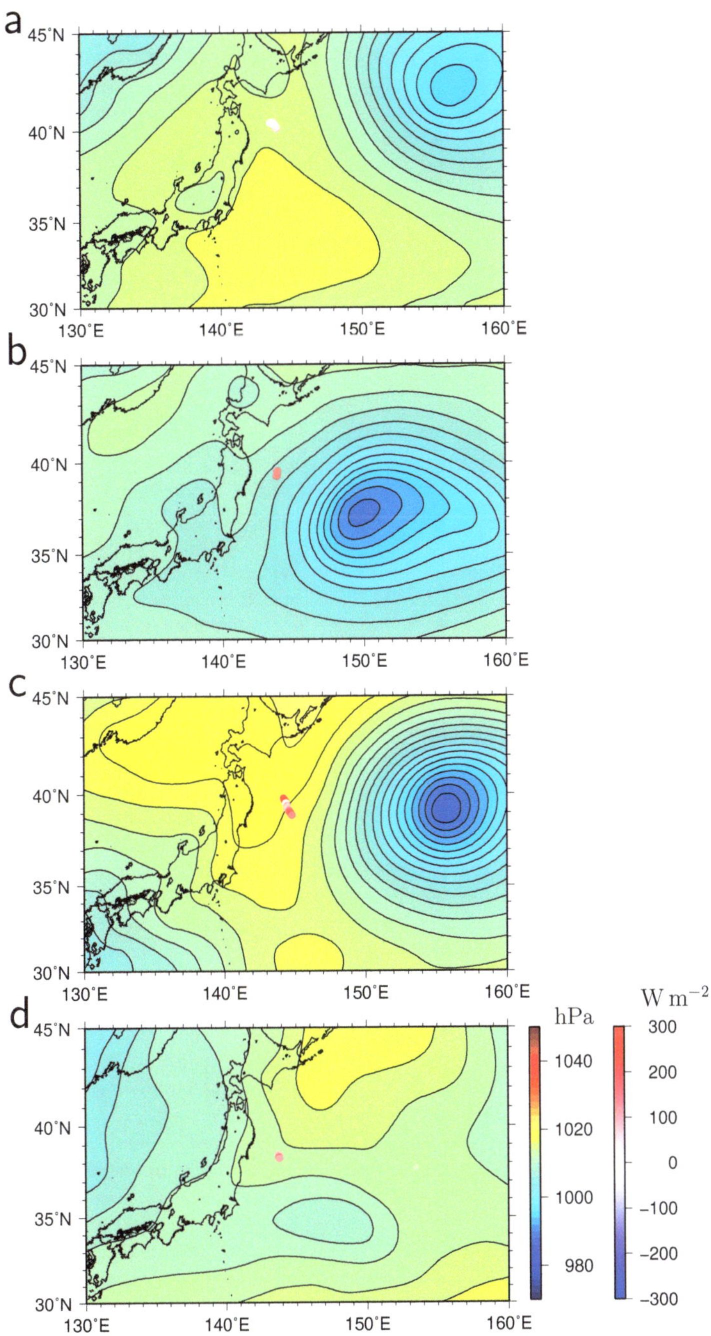


**Figure 9.** Maps of sea-level pressure (hPa) on (**a**) 24 May, 15:00, (**b**) 7 June, 21:00, (**c**) 13 June, 21:00, and (**d**) 23 June, 03:00, 2022 around the positive peaks of the turbulent (latent plus sensible) heat flux. The contour intervals of sea-level pressure are 2 hPa. The WG tracks around the heat flux peak periods are shown by lines colored by values of heat flux (W m$^{-2}$).

## 4. Summary and Conclusions

In the region off the Sanriku coast of Japan, also known as the perturbed waters from the Kuroshio and Oyashio, are distributed in a convoluted manner. In addition to the mesoscale WCRs shed by the KE, smaller scale (sub-mesoscale) warm water regions exist in the perturbed area; however, these such sub-mesoscale features cannot be detected clearly by satellites. The very cold dry wind from the Eurasian Continent is known to deprive a huge amount of heat from the Kuroshio water, which contributes to the global heat balance. Heat fluxes over such sub-mesoscale warm waters in other seasons may affect Japan's regional climate and the weather. In this study, we focused on the atmosphere and ocean in late spring, the humidity variable season. Due to the potential for research vessels to affect the ocean and atmosphere boundary layers, smaller platforms are preferred for estimating turbulent heat flux. Therefore, we observed the atmosphere and ocean using a wave-propelled autonomous surface vehicle called a Wave Glider that was developed by Liquid Robotics.

In addition to meteorological sensors installed by default for measuring the wind speed/direction and air temperature, we equipped the WG with CTD JES10mini and HOBO air temperature/relative humidity sensors. By covering the HOBO sensors with Gore-Tex fabric, we prevented the sensors from being damaged by seawater immersion. The WG was deployed on 11 May 2022 and was recovered on 5 July 2022. As a result, we were able to measure atmospheric and oceanic data for 55 days despite the severe weather conditions of several atmospheric low-pressure system crossings. The variation amplitude in the air temperature detected by the HOBO sensor covered by the Gore-Tex fabric was found to be 77% of that observed by the Weather Station sensor. Therefore, the variation in the specific humidity measured by the HOBO sensor may be underestimated by approximately 33%.

Throughout the observation period, the air temperature increased with time along with the seasonal increase in SST. Along the northwestern rim of the North Pacific subtropical high, a southerly wind transported moist air, resulting in a rise in the specific humidity prior to and in the Baiu season. Beyond seasonal variations, the WG also detected signals of different origin. Particularly, the WG reported an SST increase of more than 5 °C when passing over the Oyashio SST front. Because of the increase in SST, the air temperature and specific humidity also increased. After entering the warm Kuroshio water zone, the WG detected warmer and saltier sub-mesoscale water regions. Cold dry air masses advected by northerly winds after the passage of atmospheric low-pressure systems were observed as the air temperature and specific humidity decreased.

We estimated LHF and SHF from the atmospheric and oceanic variables obtained from the WG observation using the bulk flux algorithm. The heat flux variations were mostly caused by variations in air temperature and specific humidity rather than variations in wind speed. Although the humidity variation is possibly underestimated, we obtained the turbulent heat flux variation due to the humidity variation over ocean mesoscale and sub-mesoscale warm waters in the perturbed area. The LHF and SHF were slightly downward in the low SST region north of the Oyashio front. Substantially upward LHF and SHF were induced as cold dry air masses were advected to over sub-mesoscale warm water regions, primarily by northerly winds following the passage of atmospheric disturbances, whereas no significant heat flux was produced in the surrounding Kuroshio water region.

These results imply that heat flux studies only based on the satellite data possibly underestimate the impacts of ocean sub-mesoscale variations on the atmosphere because turbulent heat fluxes due to sub-mesoscale variations cannot be monitored by satellite data because of the low spatial resolution. In addition to wide-coverage observations by satellites, high-resolution observations by USVs are required in the future to obtain more exact knowledge on the role of the oceanic mesoscale and sub-mesoscale variations in the regional climate and the weather and are also useful to improve climate and weather forecasts.

**Author Contributions:** Conceptualization, A.N., T.H. and K.A.; methodology, T.F., N.F. and F.T.; investigation, T.F. and N.F.; validation, N.F.; formal analysis, A.N.; writing—original draft preparation, A.N.; writing—review and editing, T.H. and K.A.; project administration, T.I., K.A. and R.H.; funding acquisition, A.N. and K.A. All authors have read and agreed to the published version of the manuscript.

**Funding:** This work was partly supported by the Japan Society for the Promotion of Sci ence, Grant-in-Aid for Scientific Research (Grant Numbers: JP15H04228, JP17K05660, JP19H05596, JP20K04072, JP20H02236, and JP20KK0097).

**Institutional Review Board Statement:** Not applicable.

**Informed Consent Statement:** Not applicable.

**Data Availability Statement:** NOAA OI SST V2 High-Resolution Dataset was obtained from the website of NOAA Physical Science Laboratory (https://psl.noaa.gov/data/gridded/index.html, accessed on 9 December 2022). JRA-55 sea-level pressure data were provided by Center for Computational Sciences, University of Tsukuba (http://gpvjma.ccs.hpcc.jp/~jra55/index.html, accessed on 9 December 2022).

**Acknowledgments:** The authors thank the captain, crew, and technicians (Kaiyo Engineering Co., Ltd.) on board the cruises. The authors thank Kensuke Watari (Offshore Technologies) for his help in installing the JES10mini to the Wave Glider and Enago (www.enago.jp, accessed on 9 December 2022) for the English language review. Additionally, the authors would like to offer our special thanks to the Guest Editor (Federico Angelini) for inviting us to this special session and the reviewers for helpful comments.

**Conflicts of Interest:** The authors declare no conflict of interest.

## Abbreviations

The following abbreviations are used in this manuscript:

| | |
|---|---|
| CTD | Conductivity-temperature-depth |
| AP | Air pressure |
| AT | Air temperature |
| FSR | Full-scale range |
| GNSS | Global navigation satellite system |
| JRA-55 | Japanese 55-year Reanalysis |
| JST | Japanese standard time |
| KE | Kuroshio Extension |
| LHF | Latent heat flux |
| NOAA | National Oceanic and Atmospheric Administration |
| OI | Optimal interpolation |
| RH | Relative humidity |
| R/V | Research vessel |
| SH | Specific humidity |
| SHF | Sensible heat flux |
| SSS | Sea surface salinity |
| SST | Sea surface temperature |
| USV | Unmanned surface vehicle |
| WCR | Warm-core ring |
| WG | Wave Glider |

## References

1. Bryden, H.; Imawaki, S. Ocean heat transport. In *Ocean Circulation and Climate: Observing and Modeling the Global Ocean*; Siedler, H., Church, J., Gould, J., Eds.; Academic Press: London, UK, 2001; pp. 455–474.
2. Nagano, A.; Ichikawa, K.; Ichikawa, H.; Konda, M.; Murakami, K. Synoptic flow structures in the confluence region of the Kuroshio and the Ryukyu Current. *J. Geophys. Res.* **2009**, *114*. [CrossRef]
3. Nagano, A.; Ichikawa, K.; Ichikawa, H.; Tomita, H.; Tokinaga, H.; Konda, M. Stable volume and heat transports of the North Pacific subtropical gyre revealed by identifying the Kuroshio in synoptic hydrography south of Japan. *J. Geophys. Res.* **2010**, *115*. [CrossRef]

4. Nitani, H. Beginning of the Kuroshio. In *Kuroshio-Its Physical Aspects*; Stommel, H.; Yoshida, K., Eds.; University of Tokyo Press: Tokyo, Japan, 1972; pp. 129–163.
5. Konda, M.; Ichikawa, H.; Tomita, H.; Cronin, M.F. Surface heat flux variations across the Kuroshio Extension as observed by surface flux buoys. *J. Clim.* **2010**, *23*, 5206–5221. [CrossRef]
6. Nagano, A.; Ando, K. Saildrone-observed atmospheric boundary layer response to winter mesoscale warm spot along the Kuroshio south of Japan. *Prog. Earth Planet. Sci.* **2020**, *7*, 43. [CrossRef]
7. Kawai, H. Hydrography of the Kuroshio Extension. In *Kuroshio-Its Physical Aspects*; Stommel, H., Yoshida, K., Eds.; University of Tokyo Press: Tokyo, Japan, 1972; pp. 235–352.
8. Tomita, H.; Hihara, T.; Kako, S.; Kubota, M.; Kutsuwada, K. An introduction to J-OFURO3, a third-generation Japanese ocean flux data set using remote-sensing observations. *J. Oceanogr.* **2019**, *75*, 171–194. [CrossRef]
9. Manley, J.E. Unmanned surface vehicles, 15 years of development. In Proceedings of the OCEANS 2008, Quebec City, QC, Canada, 15–18 September 2008; pp. 1–4. [CrossRef]
10. Nagano, A.; Geng, B.; Richards, K.J.; Cronin, M.F.; Taniguchi, K.; Katsumata, M.; Ueki, I. Coupled atmosphere-ocean variations on timescales of days observed in the western tropical Pacific warm pool during mid-March 2020. *J. Geophys. Res.* **2022**, *127*, e2022JC019032. [CrossRef]
11. Kondo, J. Air-sea bulk transfer coefficients in diabatic conditions. *Boundary-Layer Meteorol.* **1975**, *9*, 91–112. [CrossRef]
12. Fairall, C.W.; Bradley, E.F.; Hare, J.E.; Grachev, A.A.; Edson, J.B. Bulk Parameterization of Air–sea Fluxes: Updates and Verification for the COARE Algorithm. *J. Clim.* **2003**, *16*, 571–591. [CrossRef]
13. Iinuma, T.; Kido, M.; Ohta, Y.; Fukuda, T.; Tomita, F.; Ueki, I. GNSS-acoustic Observations of Seafloor Crustal Deformation Using a Wave Glider. *Front. Earth Sci.* **2021**, *9*, 600946. [CrossRef]
14. Hine, R.; Wilcox, S.; Hine, G.; Richardson, T. The Wave Glider: A Wave-powered Autonomous Marine Vehicle. In Proceedings of the MTS/IEEE OCEANS 2009, Biloxi, MI, USA, 26–29 October 2009; pp. 1–6. [CrossRef]
15. Grare, L.; Statom, N.M.; Pizzo, N.; Lenain, L. Instrumented Wave Gliders for Air-sea Interaction and Upper Ocean Research. *Front. Mar. Sci.* **2021**, *8*, 664728. [CrossRef]
16. Gibson, P.; Rivin, D.; Berezin, A.; Nadezhdinskii, A. *Measurement of Water Vapor Diffusion Through Laminated Fabrics and Membranes Using a Diode Laser Spectroscope*; Technical Report NATICK/IR-98/025; U.S. Army Natock Research Development and Engineering Center: Natick, MA, USA, 1998.
17. Reynolds, R.W.; Smith, T.M.; Liu, C.; Chelton, D.B.; Casey, K.S.; Schlax, M.G. Daily High-Resolution-Blended Analyses for Sea Surface Temperature. *J. Clim.* **2007**, *20*, 5473–5496. [CrossRef]
18. Kobayashi, S.; Ota, Y.; Harada, Y.; Ebita, A.; Moriya, M.; Onoda, H.; Onogi, K.; Kamahori, H.; Kobayashi, C.; Endo, H.; et al. The JRA-55 Reanalysis: General Specifications and Basic Characteristics. *J. Meteor. Soc. Jpn.* **2015**, *93*, 5–48. [CrossRef]
19. Tomita, T.; Yamaura, T.; Hashimoto, T. Interannual Variability of the Baiu Season near Japan Evaluated from the Equivalent Potential Temperature. *J. Meteor. Soc. Jpn.* **2011**, *89*, 517–537. [CrossRef]
20. Hasegawa, T.; Nagano, A.; Matsumoto, H.; Ariyoshi, K.; Wakita, M. El Niño-related sea surface elevation and ocean bottom pressure enhancement associated with the retreat of the Oyashio southeast of Hokkaido, Japan. *Mar. Geophys. Res.* **2019**, *40*, 505–512. [CrossRef]

sensors

MDPI

Article

# Optimization of the NRCS Sampling at the Sea Wind Retrieval by the Airborne Rotating-Beam Scatterometer Mounted under Fuselage

**Alexey Nekrasov [1,2,*], Alena Khachaturian [1] and Evgenii Vorobev [1]**

1 Department of Radio Engineering Systems, Saint Petersburg Electrotechnical University, Professora Popova 5F, 197022 Saint Petersburg, Russia; khachaturyan.al@gmail.com (A.K.); vorobuevgeniy@gmail.com (E.V.)
2 Institute for Computer Technologies and Information Security, Southern Federal University, Chekhova 2, 347922 Taganrog, Russia
* Correspondence: alexei-nekrassov@mail.ru; Tel.: +7-8634-360-484

**Abstract:** The optimization of normalized radar cross-section (NRCS) sampling by a scatterometer allows an increase in the accuracy of the wind retrieval over the water surface and a decrease in the time of the measurement. Here, we investigate the possibility of improving wind vector measurement with an airborne rotating-beam scatterometer mounted under the fuselage. For this purpose, we investigated NRCS sampling at various incidence angles, and the possibility of using NRCS samples obtained during simultaneous measurement at different incidence angles to perform wind retrieval. The proposed wind algorithms are based on a geophysical model function (GMF). Sea wind retrieval was carried out using Monte Carlo simulations with consideration of a single incidence angle or combinations of several incidence angles. The incidence angles of interest were 30°, 35°, 40°, 45°, 50°, 55°, and 60°. The simulation showed that the wind speed error decreased with an increase in the incidence angle, and the wind direction error tended to decrease with an increase in the incidence angle. The single incidence angle case is characterized by higher maximum wind retrieval errors but allows for a higher maximum altitude of the wind retrieval method's applicability to be achieved. The use of several neighboring incidence angles allows a better wind vector retrieval accuracy to be achieved. The combinations of three and four incidence angles provided the lowest maximum wind speed and direction errors in the range of the incidence angles from 45° to 60° but, unfortunately, provide the lowest maximum altitude of applicability of the wind retrieval method. At the same time, the combination of two incidence angles is characterized by slightly higher maximum wind retrieval errors than in the cases of three and four incidence angles, but they are lower than in the case of the single incidence angle. Moreover, the two incidence angles' combination is a simpler way to decrease the wind retrieval errors, especially for measurement near an incidence angle of 30°, providing nearly the highest maximum altitude of the wind retrieval method applicability. The results obtained can be used to enhance existing airborne radars and in the development of new remote sensing systems.

**Keywords:** radar; airborne scatterometer; radar backscatter; sea surface; sea wind retrieval

## 1. Introduction

During the last decades, sea-surface backscattering has been of great interest to researchers. This interest is motivated by the need for a better understanding of sea-surface backscattering as a physical phenomenon and by its prospective application in the development and improvement of remote sensing technology. Therefore, research on sea-surface backscattering is very important to understanding the formation mechanism of sea clutter, which is crucial for radar target detection in nonhomogeneous environments [1–4], and for operational monitoring of waves, currents, and sea winds [5–7].

Water backscattering is studied by means of a sensor called a scatterometer. Experiments have been performed in wind-wave tanks [7,8], on sea platforms [9,10], and by airborne [11,12] and spaceborne [13,14] scatterometers. The identified relationship between the backscatter and wind vector over sea made it possible to use scatterometers for remote measurement of the wind vector over water surfaces [15].

Near-surface wind retrieval is performed with a wind algorithm. The wind algorithm is based on a GMF and takes into account the specificity of the measuring geometry of a scatterometer [16].

Scatterometers placed on one or several satellites provide current information about the wind conditions over oceans and seas at a global scale. At the same time, scatterometers' placement on aircraft allows local information on the wind over water to be obtained, which can clarify the information received from satellites for meteorological and navigation applications and for scientific purposes.

Airborne scatterometers (or multimode radars with a scatterometer mode) have a fixed-beam antenna [17–20], scanning antenna [21–23], or rotating-beam antenna [12,24–28]. Mostly, antennas rotating in the horizontal plane are installed on the bottom or under the fuselage.

Scatterometers with a fixed-beam antenna require the measurements to be on a circular ground track [19,20,29]. On the contrary, scatterometers with scanning [30–32] or rotating-beam [12,27] antennas require the measurements to be on a rectilinear ground track.

Airborne sea-wind measurements using rotating-beam scatterometers has quite a long heritage. The prime examples of such scatterometers are KU-SCAT (Ku-band scatterometer) and C-SCAT (C-band scatterometer) of the Microwave Remote Sensing Laboratory at the University of Massachusetts Amherst [12], DUTSCAT (multifrequency Delft University of Technology Scatterometer) [33], RACS (German Rotating Antenna C-band Scatterometer) [11], IWRAP (C- and Ka-band Imaging Wind and Rain Airborne Profiler) of the Microwave Remote Sensing Laboratory at the University of Massachusetts Amherst [24], and DopplerScatt (Ka-band pencil-beam Doppler scatterometer) of the NASA Instrument Incubator Program [6].

In the case of airborne scatterometers, the rotating antenna has one or several pencil beams (Figure 1), or a fan beam (Figure 2) [24,34,35]. Multiple beams located in the same vertical plane at different incidence angles allow the measured NRCSs (simultaneously) to be obtained at all their incidence angles. A similar capability is shown by the fan beam when time-delay selection is applied.

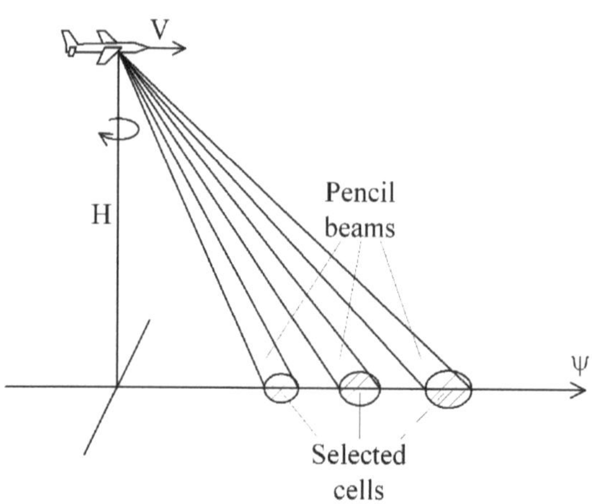


**Figure 1.** Rotating antenna multibeam geometry (three-beam case in the vertical plane): $V$ is the speed of flight; $H$ is the altitude; $\psi$ is the aircraft flight direction.

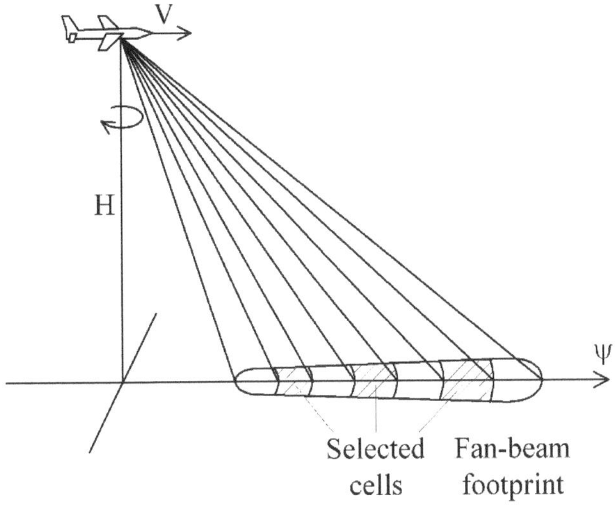


**Figure 2.** Rotating antenna fan-beam geometry (three selected sell case in the vertical plane): $V$ is the speed of flight; $H$ is the altitude; $\psi$ is the aircraft flight direction.

Usually, only one incidence angle is used for wind retrieval by an airborne scatterometer with a rotating antenna. However, a multibeam or fan-beam geometry can achieve NRCS sampling at several incidence angles in the same vertical plane. In this connection, this study was motivated by the need for an enhancement in the functionality of radars with such observation geometries and further increases in the wind retrieval accuracy. The simultaneous use of the measured NRCSs at several incidence angles in the same plane seems promising for wind measurement by airborne scatterometers (or multimode radars with the scatterometer mode) mounted under the fuselage. Thus, this manuscript addresses the analysis of such geometries and their possible implementation for better wind retrieval over the sea, e.g., with airborne scatterometers or enhanced airborne maritime/ground surveillance radars.

Section 2 introduces the background of wind retrieval using a scatterometer and the wind retrieval algorithms developed to estimate the wind vector over the sea by airborne scatterometers with a rotating antenna sampling NRCSs at a single incidence angle or combinations of several incidence angles. Section 3 describes the simulations, presents the results obtained and their discussion, and suggestions for future research. Finally, the conclusions are outlined in Section 4.

## 2. Materials and Methods

A wind scatterometer is an airborne or spaceborne microwave sensor designed for operational measurement of the wind vector over the ocean or sea [7]. The wind vector retrieval by a scatterometer depends on NRCSs sampling from different azimuthal directions (and different or the same incidence angles depending on the scatterometer configuration and its installation on an aircraft or satellite) and a water GMF representing the NRCS $\sigma^\circ(U,\theta,\alpha)$ dependence on the wind speed $U$, incidence angle $\theta$, and azimuthal angle $\alpha$ relative to the up-wind direction. The GMFs are described in various analytical forms and can be presented only as a table. One such analytical GMF form is as follows [36]:

$$\sigma^\circ(U,\theta,\alpha) = A(U,\theta) + B(U,\theta)\cos\alpha + C(U,\theta)\cos(2\alpha), \tag{1}$$

where $A(U,\theta)$, $B(U,\theta)$, and $C(U,\theta)$ are the coefficients written as $A(U,\theta) = a_0(\theta)U^{\gamma_0(\theta)}$, $B(U,\theta) = a_1(\theta)U^{\gamma_1(\theta)}$, and $C(U,\theta) = a_2(\theta)U^{\gamma_2(\theta)}$; $a_0(\theta)$, $a_1(\theta)$, $a_2(\theta)$, $\gamma_0(\theta)$, $\gamma_1(\theta)$, and

$\gamma_2(\theta)$ are the coefficients corresponding to the appropriate incidence angle, radar wavelength, and polarization.

In the general case, wind vector retrieval by an airborne scatterometer with a rotating antenna that samples NRCSs at one incidence angle only can be achieved by solving the system of $N$ equations [28,37]:

$$\begin{cases} \sigma^\circ(U,\theta,\alpha+\psi_1) = A(U,\theta) + B(U,\theta)\cos(\alpha+\psi_1) + C(U,\theta)\cos(2(\alpha+\psi_1)), \\ \cdots\cdots\cdots\cdots\cdots\cdots\cdots\cdots\cdots\cdots\cdots\cdots \\ \sigma^\circ(U,\theta,\alpha+\psi_i) = A(U,\theta) + B(U,\theta)\cos(\alpha+\psi_i) + C(U,\theta)\cos(2(\alpha+\psi_i)), \\ \cdots\cdots\cdots\cdots\cdots\cdots\cdots\cdots\cdots\cdots\cdots\cdots \\ \sigma^\circ(U,\theta,\alpha+\psi_N) = A(U,\theta) + B(U,\theta)\cos(\alpha+\psi_N) + C(U,\theta)\cos(2(\alpha+\psi_N)), \end{cases} \quad (2)$$

where $i = \overrightarrow{1,N}$, $N$ is the number of the azimuth sectors observed during a whole 360° azimuth observation, $N = 360°/\Delta\alpha_s$; $\Delta\alpha_s$ is the angular width of each azimuth sector (composing whole 360° azimuth NRCS curve); $\sigma^\circ(U,\theta,\alpha+\psi_i)$ is the measured NRCS corresponding to the azimuth sector number $I$; and $\psi_i$ is the direction of the azimuth sector number $i$ relative to the aircraft flight direction $\psi$. The system of Equation (2) or similar systems of equations for wind retrieval over water are composed based on GMF Equation (1) under the condition of a narrow antenna beam in the azimuth plane, where the azimuth sector angular width is 15–20° [38,39].

As the system of Equation (2) provides the up-wind direction retrieval, it is converted to the measured wind direction $\psi_w$ using the following equation [40]:

$$\psi_w = \psi - \alpha \pm 180°. \quad (3)$$

In the case of an airborne scatterometer with a multibeam or fan-beam rotating antenna, it can provide simultaneous NRCS sampling at several incidence angles in the same vertical plane, which seems more advantageous compared to NRCS sampling at only one incidence angle. Thus, the following system of equations can be used for wind retrieval:

$$\begin{cases} \sigma^\circ(U,\theta_1,\alpha+\psi_1) = A(U,\theta_1) + B(U,\theta_1)\cos(\alpha+\psi_1) + C(U,\theta_1)\cos(2(\alpha+\psi_1)), \\ \cdots\cdots\cdots\cdots\cdots\cdots\cdots\cdots\cdots\cdots\cdots\cdots \\ \sigma^\circ(U,\theta_1,\alpha+\psi_i) = A(U,\theta_1) + B(U,\theta_1)\cos(\alpha+\psi_i) + C(U,\theta_1)\cos(2(\alpha+\psi_i)), \\ \cdots\cdots\cdots\cdots\cdots\cdots\cdots\cdots\cdots\cdots\cdots\cdots \\ \sigma^\circ(U,\theta_1,\alpha+\psi_N) = A(U,\theta_1) + B(U,\theta_1)\cos(\alpha+\psi_N) + C(U,\theta_1)\cos(2(\alpha+\psi_N)), \\ \cdots\cdots\cdots\cdots\cdots\cdots\cdots\cdots\cdots\cdots\cdots\cdots \\ \sigma^\circ\left(U,\theta_j,\alpha+\psi_1\right) = A\left(U,\theta_j\right) + B\left(U,\theta_j\right)\cos(\alpha+\psi_1) + C\left(U,\theta_j\right)\cos(2(\alpha+\psi_1)), \\ \cdots\cdots\cdots\cdots\cdots\cdots\cdots\cdots\cdots\cdots\cdots\cdots \\ \sigma^\circ\left(U,\theta_j,\alpha+\psi_i\right) = A\left(U,\theta_j\right) + B\left(U,\theta_j\right)\cos(\alpha+\psi_i) + C\left(U,\theta_j\right)\cos(2(\alpha+\psi_i)), \\ \cdots\cdots\cdots\cdots\cdots\cdots\cdots\cdots\cdots\cdots\cdots\cdots \\ \sigma^\circ\left(U,\theta_j,\alpha+\psi_N\right) = A\left(U,\theta_j\right) + B\left(U,\theta_j\right)\cos(\alpha+\psi_N) + C\left(U,\theta_j\right)\cos(2(\alpha+\psi_N)), \\ \cdots\cdots\cdots\cdots\cdots\cdots\cdots\cdots\cdots\cdots\cdots\cdots \\ \sigma^\circ(U,\theta_K,\alpha+\psi_1) = A(U,\theta_K) + B(U,\theta_K)\cos(\alpha+\psi_1) + C(U,\theta_K)\cos(2(\alpha+\psi_1)), \\ \cdots\cdots\cdots\cdots\cdots\cdots\cdots\cdots\cdots\cdots\cdots\cdots \\ \sigma^\circ(U,\theta_K,\alpha+\psi_i) = A(U,\theta_K) + B(U,\theta_K)\cos(\alpha+\psi_i) + C(U,\theta_K)\cos(2(\alpha+\psi_i)), \\ \cdots\cdots\cdots\cdots\cdots\cdots\cdots\cdots\cdots\cdots\cdots\cdots \\ \sigma^\circ(U,\theta_K,\alpha+\psi_N) = A(U,\theta_K) + B(U,\theta_K)\cos(\alpha+\psi_N) + C(U,\theta_K)\cos(2(\alpha+\psi_N)), \end{cases} \quad (4)$$

where $j = \overrightarrow{1,K}$, $K$ is the number of the incidence angles observed (or used for wind retrieval in the case of a multibeam or fan-beam antenna rotating in the horizontal plane), $\sigma^\circ(U,\theta_j,\alpha+\psi_i)$ is the measured NRCS corresponding to incidence angle number $j$, and azimuth sector number $i$. The system of Equation (4) is also composed based on GMF

Equation (1) for each azimuth sector observed during the whole 360° azimuth observation at each incidence angle of interest under the conditions of a narrow antenna beam in the azimuth plane, where the azimuth sector angular width is 15–20° [38,39].

The GMF form of Equation (1) has a particular feature in that the azimuthally averaged NRCS at the same incidence angle $\sigma^{\circ}_{av360^{\circ}}(U,\theta)$ can be written as [41]:

$$\sigma^{\circ}_{av360^{\circ}}(U,\theta) = \frac{1}{N}\sum_{i=1}^{N}\sigma^{\circ}(U,\theta,\alpha+\psi_i) = A(U,\theta) = a_0(\theta)U^{\gamma_0(\theta)}, \tag{5}$$

and this feature can be applied to simplify and speed up the wind speed estimation procedure using a modified system of the equation obtained from the system of Equation (4) with the help of Equation (5):

$$\begin{cases} \sigma^{\circ}_{av360^{\circ}}(U,\theta_1) = a_0(\theta_1)U^{\gamma_0(\theta_1)}, \\ \dots\dots\dots\dots\dots\dots\dots\dots \\ \sigma^{\circ}_{av360^{\circ}}(U,\theta_j) = a_0(\theta_j)U^{\gamma_0(\theta_j)} \\ \dots\dots\dots\dots\dots\dots\dots\dots \\ \sigma^{\circ}_{av360^{\circ}}(U,\theta_K) = a_0(\theta_K)U^{\gamma_0(\theta_K)}. \end{cases} \tag{6}$$

Then, the wind direction is calculated using the system of Equations (3) and (4).

Thus, in the case of an airborne rotating-antenna scatterometer with a multibeam or fan-beam antenna geometry installed at the bottom or under an aircraft, simultaneous NRCS sampling at several incidence angles can be used to recover the wind vector over water surfaces.

## 3. Results and Discussion

To evaluate the proposed wind retrieval algorithm and optimize the wind retrieval procedure, we investigated NRCS sampling at various incidence angles and the possibility of using the NRCS samples obtained during simultaneous measurements at different incidence angles with the help of the wind algorithm proposed in Section 2.

For this purpose, we completed Monte Carlo simulations using a Rayleigh power (exponential) distribution and a GMF from Equation (1), with the Ku-band coefficients corresponding to the horizontal polarization [42]:

$$\begin{aligned} &a_0(\theta) = 10^{2.47324-0.224780\theta+0.001499\theta^2},\ a_1(\theta) = 10^{-0.50593-0.11694\theta+0.000484\theta^2}, \\ &a_2(\theta) = 10^{1.63685-0.2100488\theta+0.001383\theta^2},\ \gamma_0(\theta) = -0.15+0.071\theta-0.0004\theta^2, \\ &\gamma_1(\theta) = -0.02+0.061\theta-0.0003\theta^2,\ \gamma_2(\theta) = -0.16+0.074\theta-0.0004\theta^2. \end{aligned} \tag{7}$$

The incidence angles of interest were 30°, 35°, 40°, 45°, 50°, 55°, and 60°. The whole 360° azimuth circles observed were divided into $N = 72$ azimuth sectors, which provided an azimuth sector width of 5°. In total, 87 "measured" NRCS samples, under the assumption of a 0.2 dB instrumental noise, were generated for each azimuthal sector and each incidence angle of interest. Wind retrieval was performed at wind speeds of 2 to 30 m/s during various combinations of the incidence angles to evaluate their potential and the accuracy of wind vector retrieval. For each combination of wind speed and azimuth angle at each incidence angle of interest, 30 independent trials were performed.

First, we evaluated the maximum errors of the wind speed and direction retrieval when only one incidence angle was used. The system of Equation (2) was used for this purpose in the simulation. These simulation results are presented in Appendix A (Figures A1–A7, respectively, for the incidence angles of 30°, 35°, 40°, 45°, 50°, 55°, and 60°). The wind retrieval maximum errors were 0.73 m/s and 5.6° at $\theta$ = 30°, 0.7 m/s and 5.2° at $\theta$ = 35°, 0.64 m/s and 4.5° at $\theta$ = 40°, 0.58 m/s and 4.6° at $\theta$ = 45°, 0.53 m/s and 3.8° at $\theta$ = 50°, 0.52 m/s and 4.7° at $\theta$ = 55°, and 0.51 m/s and 4.0° at $\theta$ = 60°, respectively. The comparative results are shown in Figure 3. They demonstrate that the maximum wind speed error decreased with an increase in the incidence angle. The maximum wind direction error also tended to decrease with an increase in the incidence angle.

It was expected that the higher number of incidence angles used at the wind retrieval should decrease the wind retrieval errors as the whole number of NRCS samples would be available in this case compared with the case when only one incidence angle was used at the wind retrieval. Therefore, we considered wind retrieval in other cases when the measured NRCSs at several incidence angles in the same plane were used simultaneously. The simulations of these cases were performed using the system of Equation (4).

The simulation results in the case of two neighboring incidence angles used for wind retrieval are presented in Appendix B (Figures A8–A13, respectively, for the combinations of the incidence angles of 30° and 35°; 35° and 40°; 40° and 45°; 45° and 50°; 50° and 55°; and 55° and 60°. The maximum errors of the wind estimation in cases of two incidence angles are 0.53 m/s and 5.2° at $\theta$ = (30°, 35°), 0.54 m/s and 4.7° at $\theta$ = (35°, 40°), 0.47 m/s and 3.8° at $\theta$ = (40°, 45°), 0.42 m/s and 3.3° at $\theta$ = (45°, 50°), 0.41 m/s and 3.2° at $\theta$ = (50°, 55°), and 0.36 m/s and 3.6° at $\theta$ = (55°, 60°), respectively.

The results obtained in the case of three neighboring incidence angles for wind retrieval are presented in Appendix C (Figures A14–A18, respectively, for the combinations of the incidence angles of 30°, 35°, and 40°; 35°, 40°, and 45°; 40°, 45°, and 50°; 45°, 50°, and 55°; and 50°, 55°, and 60°). The maximum errors of the wind speed and direction retrieval in cases of three incidence angles are 0.49 m/s and 5.1° at $\theta$ = (30°, 35°, 40°), 0.54 m/s and 4.7° at $\theta$ = (35°, 40°, 45°), 0.41 m/s and 3.5° at $\theta$ = (40°, 45°, 50°), 0.34 m/s and 3.2° at $\theta$ = (45°, 50°, 55°), and 0.34 m/s and 3.0° at $\theta$ = (50°, 55°, 60°), respectively.

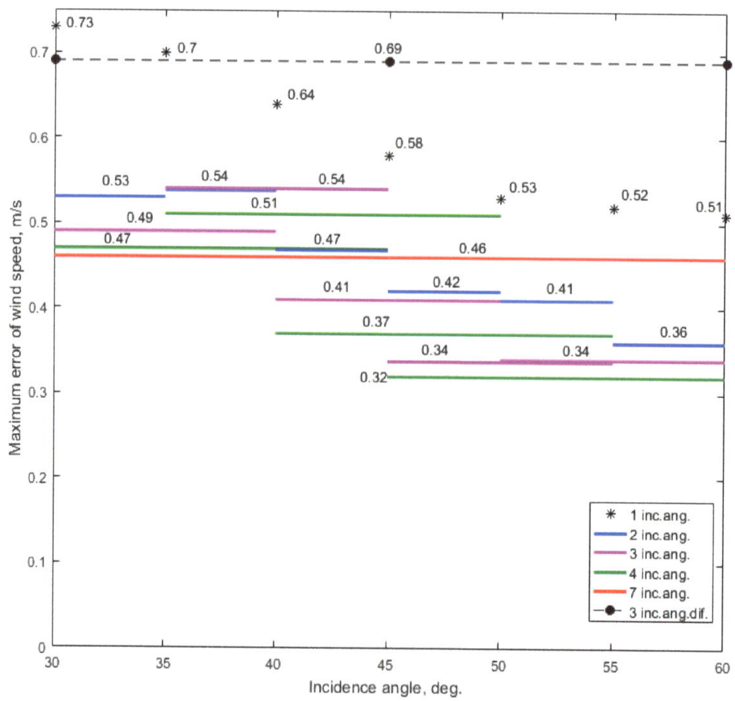


(a)

**Figure 3.** *Cont.*

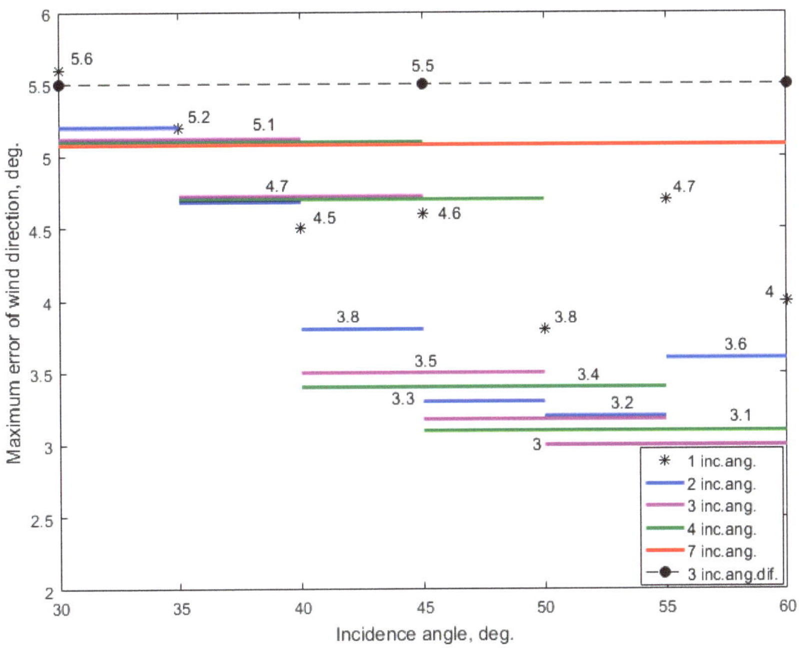


(b)

**Figure 3.** Comparative results for the maximum wind retrieval errors in accordance with the cases considered: (**a**) maximum error of the wind speed; (**b**) maximum error of the wind direction. The black asterisks represent the wind retrieval when one incidence angle was used; the blue lines represent the wind retrieval when two incidence angles were used; the purple lines represent the wind retrieval when three incidence angles were used; the green lines represent the wind retrieval when four incidence angles were used; the red line represents the wind retrieval when seven incidence angles were used; the black dashed line with dots represents the wind retrieval when three incidence angles were used but with a 15° incidence angle difference between the neighboring incidence angles in the range of considered incidence angles of 30° to 60°.

The simulation results of when four neighboring incidence angles were used for wind retrieval are presented in Appendix D (Figures A19–A22, respectively, for the combinations of the incidence angles of 30°, 35°, 40°, and 45°; 35°, 40°, 45°, and 50°; 40°, 45°, 50°, and 55°; and 45°, 50°, 55°, and 60°. The maximum errors of the wind estimation in cases of four incidence angles are 0.47 m/s and 5.1° at $\theta$ = (30°, 35°, 40°, 45°), 0.51 m/s and 4.7° at $\theta$ = (35°, 40°, 45°, 50°), 0.37 m/s and 3.4° at $\theta$ = (40°, 45°, 50°, 55°), and 0.32 m/s and 3.1° at $\theta$ = (45°, 50°, 55°, 60°), respectively.

The simulation results of when seven neighbor incidence angles were used for wind retrieval are presented in Appendix E (Figure A23 for the incidence angles' combination of 30°, 35°, 40°, 45°, 50°, 55°, and 60°. The maximum errors of the wind retrieval in the case of seven incidence angles are 0.46 m/s and 5.1° at $\theta$ = (30°, 35°, 40°, 45°, 50°, 55°, 60°).

Finally, we evaluated the maximum errors of the wind speed and direction retrieval when only three incidence angles were used but with the highest incidence angle difference of 15° between the neighboring incidence angles in the range of considered incidence angles of 30° to 60°. The simulation results are presented in Appendix F (Figure A24 for the incidence angles' combination of 30°, 45°, and 60°. The wind retrieval maximum errors in this case are 0.69 m/s and 5.5° at $\theta$ = (30°, 45°, 60°).

The summarized results presented in Figure 3 clearly demonstrate that the use of NRCSs from several neighboring incidence angles provides better accuracy of the wind speed and direction retrieval than when only one incidence angle is in use. This result, of course, was expected.

The use of NRCSs from all seven incidence angles considered (30°, 35°, 40°, 45°, 50°, 55°, 60°) provides better wind speed retrieval accuracy compared to the case of only one incidence angle. At the same time, the seven-incidence-angles case does not increase the wind direction retrieval accuracy compared to the other incidence angles and their combinations in the range of the incidence angles from 40° to 60°, providing a difference of about 2°. However, the combination of seven incidence angles is not the best solution for increasing the accuracy of wind retrieval using a rotating-beam scatterometer.

Unfortunately, the use of only three incidence angles (30°, 45°, 60°) with the highest incidence angle difference of 15° between the neighboring incidence angles in the range of incidence angles of 30° to 60° showed an even worse result compared to the combination of seven incidence angles (30°, 35°, 40°, 45°, 50°, 55°, 60°).

Figure 3 demonstrates that the application of the combinations of two, three, and four incidence angles (excluding the case of three incidence angles at $\theta$ = (30°, 45°, 60°)) reduces the error in the wind speed and direction retrieval. The lowest value of the maximum wind speed errors is achieved with the combinations of three and four incidence angles in the range of the incidence angles from 45° to 60°. The lowest value of the maximum wind direction errors also corresponds to the numbers of the combinations of incidence angles in the same range as the incidence angles.

Nevertheless, the use of the combination of two incidence angles also demonstrates good wind retrieval accuracy compared to the case of only one incidence angle, and it is slightly worse than the accuracy achieved with the combinations of three or four incidence angles. Thus, wind retrieval within the combination of two incidence angles can be used as a simpler way to increase the wind retrieval accuracy, especially when NRCS sampling is only available near an incidence angle of 30° due to the scatterometer's design features not allowing the application of combinations of three and four incidence angles, or the incidence angle limitation due to the size of the area observed.

The completed simulations proved that the wind retrieval errors in all the cases considered are within the typical accuracy of scatterometer wind retrieval of ±2 m/s and ±20° [43].

The area observed sets the maximum altitude limitation of airborne rotating-beam scatterometers' applicability, as the observation circles traced on the water surface at the used incidence angles should be within this area. It is assumed that the wind and wave conditions can be considered to be the same in all parts of the area. The maximum altitude $H_{\max}$ of the wind retrieval method's applicability for measuring such geometry is as follows:

$$H_{\max} = \frac{D_{\max}}{2\tan\theta}, \quad (8)$$

where $D_{\max}$ is the maximum diameter of the observed circular NRCS curve, which is assumed to provide the identity of the wind and wave conditions within the area of interest at the given incidence angle. For example, if the dimensions of such an area are about 15–20 km, the maximum altitudes of applicability of the considered method for the wind recovery are about 5.77 km and 17.3 km at incidence angles of 60° and 30°, respectively. Otherwise, at higher altitudes, the diameter of the observed circular NRCS curve will exceed 20 km, breaking the condition of the wind and wave identity in the observed area.

Taking this into account and applying the incidence angle step of 5° for the beams or selected cells starting with a 30° incidence angle, the maximum altitude limitations for the combinations of 1, 2, 3, and 4 incidence angles are 17.3, 14.2, 11.9, and 10 km, respectively. The lowest value of the maximum altitude limitation of 5.77 km corresponds to the case of the combination of seven incidence angles (30°, 35°, 40°, 45°, 50°, 55°, 60°) and the case of three incidence angles (30°, 45°, 60°), with the highest incidence angle difference of 15° between the neighbor incidence angles in the range of incidence angles of 30° to 60°.

Hence, the optimization of NRCS sampling during sea wind measurement using an airborne rotating-beam scatterometer mounted at the bottom or under the fuselage to increase the accuracy of the wind retrieval depends on the given altitude of measurements. If the measurement altitude requirement is only about 5.77 km, the best wind retrieval accuracy is achieved when the incidence angle or its combinations tend to the value of 60° and the combinations of three or four incident angles are used. If a higher measurement altitude is required, the incidence angle or its combinations need to be decreased properly, but this will lead to a decrease in the accuracy of the wind measurement (Figure 3). The simplest way to increase the wind measurement accuracy while providing the almost maximum altitude of measurement is to use the combination of two incident angles, as it provides lower wind speed retrieval error compared to the case of only one incidence angle, and the wind retrieval errors in the case of the combination of two incidence angles are only slightly higher than the errors generated by the use of the combinations of three or four incidence angles.

This study considered the circular NRCS sampling procedure and wind retrieval in the Ku-band. The scope of future research is the consideration of other NRCS sampling schemes in this and other bands for further improvement of the sea wind retrieval accuracy and to increase the maximum altitude of the method's applicability.

## 4. Conclusions

Analysis of wind measurement using an airborne scatterometer with a multibeam or fan-beam rotating-antenna installed at the bottom or under the fuselage showed that in the case of only one incidence angle for wind retrieval, the wind speed error decreased with an increase in the incidence angle and the wind direction error tended to decrease with an increase in the incidence angle. This case provided the highest value of the maximum altitude of the method's applicability for wind retrieval.

The use of NRCSs from several neighboring incidence angles allowed a better accuracy of the wind vector retrieval to be achieved compared to the case of only one incidence angle. The performed simulations showed that the use of the combinations of three and four incidence angles provided the lowest maximum wind speed errors in the range of incidence angles from 45° to 60°. The same result was also achieved regarding the wind direction errors of the combinations of incidence angles in this range of incidence angles. The maximum altitudes of the wind retrieval method with the combinations of three and four incidence angles were lower than in the cases of one incidence angle and two incidence angles.

At the same time, the wind retrieval errors in the case of the combination two incidence angles were only slightly higher than the errors generated with the use of the combinations of three or four incidence angles. However, in this case, the wind retrieval errors were lower than in the case of only one incidence angle. Moreover, this case can be used as a simpler way to decrease wind retrieval errors, especially for measurement near an incidence angle of 30°, when the scatterometer design features exclude the application of the combinations of three and four incidence angles, providing nearly the highest value of the maximum altitude of the applicability of the wind retrieval method.

Unfortunately, wind measurement using a rotating-beam scatterometer in the case of seven incidence angles was not the best solution to reducing wind retrieval errors. However, it provides at least a lower wind speed retrieval error compared to the case of only one incidence angle. Moreover, the combination of seven incidence angles is characterized by the lowest value of the maximum altitude of the wind retrieval method's applicability.

The combination of three incidence angles with the highest incidence angle difference of 15° between the neighboring incidence angles in the range of incidence angles of 30° to 60° also demonstrated the worst result. It provided a lower wind speed retrieval error compared to the case of only one incidence angle at 30° and 35°, and a lower wind direction retrieval error compared to the case of only one incidence angle at 30°. This case was also characterized by the lowest value of the maximum altitude of the wind retrieval method's applicability.

The errors of the wind vector retrieval with the help of the proposed wind algorithms in all considered cases of the rotating-beam scatterometers were within the ranges of a typical scatterometer's accuracy of ±2 m/s and ±20°.

Thus, the use of several neighboring incidence angles during sea-wind measurement with airborne scatterometers or multimode radars operating in the scatterometer mode provides better wind vector retrieval accuracy compared with the case of a single incidence angle. The obtained results can be used for optimization of the NRCS sampling procedure over the ocean and sea using a rotating-beam scatterometer and for the development of new sea wind sensors or enhancement of the functionality of existing airborne maritime/ground surveillance radars, extending their application possibilities to joint and standalone measurements in oceanography, meteorology, and navigation.

**Author Contributions:** Conceptualization, A.N.; methodology, A.N. and A.K.; software, A.K.; validation, A.N. and A.K.; formal analysis, A.N. and A.K.; investigation, A.N. and A.K.; resources, A.K. and E.V.; data curation, A.N., A.K. and E.V.; writing—original draft preparation, A.N.; writing—review and editing, A.N. and A.K.; visualization, A.N., A.K. and E.V.; supervision, A.N.; project administration, A.K.; funding acquisition, A.K. All authors have read and agreed to the published version of the manuscript.

**Funding:** This research was funded by the Russian Science Foundation grant number 21-79-10375, https://rscf.ru/en/project/21-79-10375/ (accessed on 5 May 2022). The APC was funded by the Russian Science Foundation grant number 21-79-10375.

**Institutional Review Board Statement:** Not applicable.

**Informed Consent Statement:** Not applicable.

**Data Availability Statement:** Data sharing not applicable.

**Conflicts of Interest:** The authors declare no conflict of interest.

## Appendix A

The simulation results of the wind retrieval with the system of Equation (2), when only one incidence angle is in use, are presented here. The results were obtained under the following conditions. The whole 360° azimuth circle observed was divided into $N = 72$ azimuth sectors (the azimuth sector width is 5°) at the directions of 0°, 5°, 10°, ..., 355° relative to the aircraft flight direction assuming 0.2 dB instrumental noise and 87 integrated NRCS samples for each azimuth sector at wind speeds of 2–30 m/s. The results obtained for the incidence angles of 30°, 35°, 40°, 45°, 50°, 55°, and 60° are presented in Figures A1–A7, respectively.

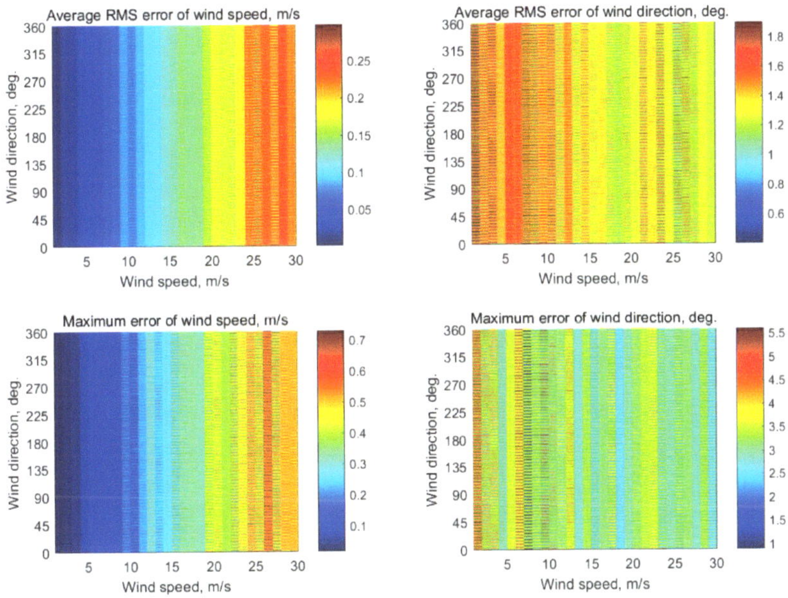


**Figure A1.** Simulation results of the use of only one incidence angle of 30° for wind retrieval.

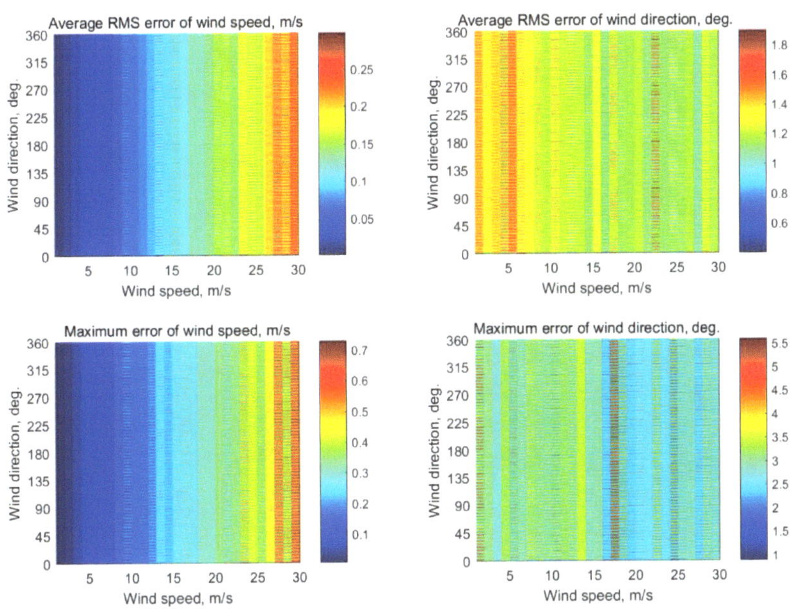


**Figure A2.** Simulation results of the use of only one incidence angle of 35° for wind retrieval.

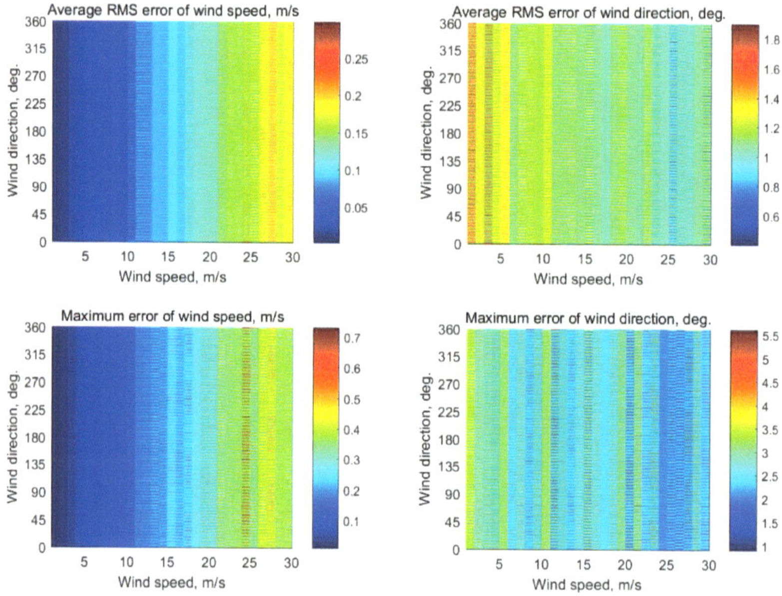


**Figure A3.** Simulation results of the use of only one incidence angle of 40° for wind retrieval.

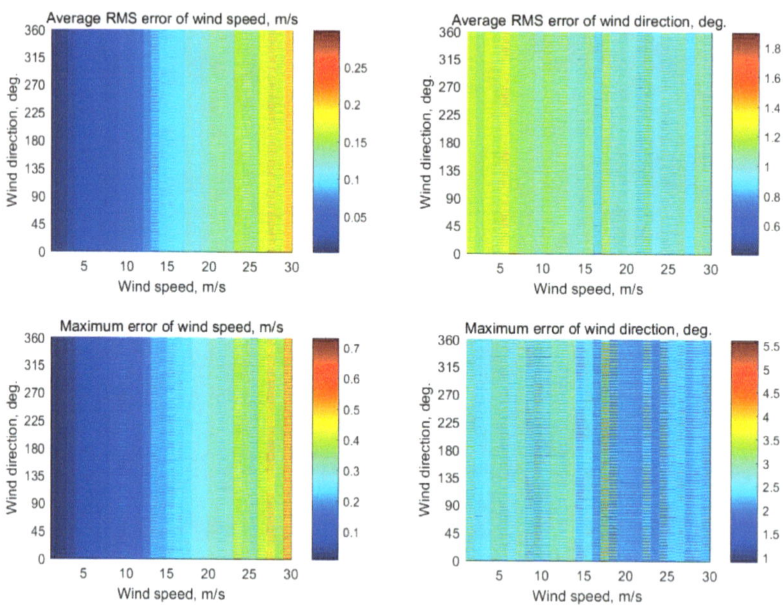


**Figure A4.** Simulation results of the use of only one incidence angle of 45° for wind retrieval.

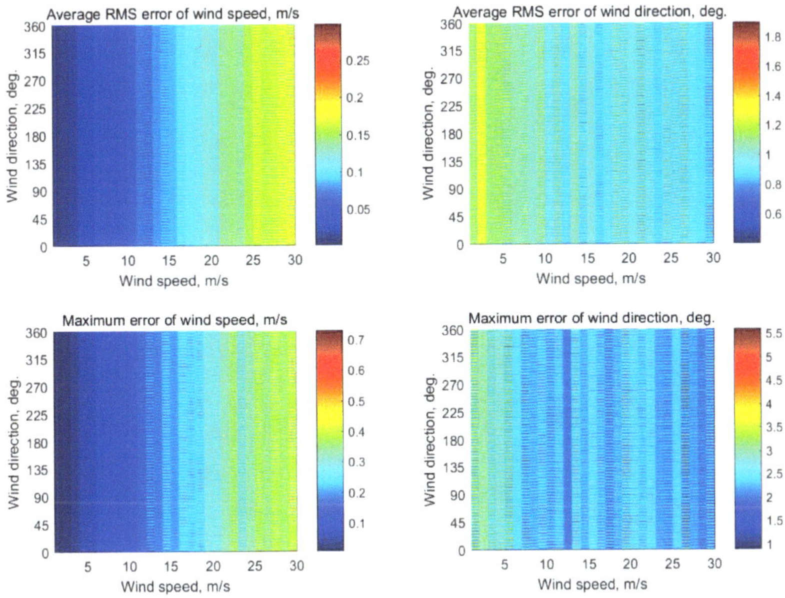


**Figure A5.** Simulation results of the use of only one incidence angle of 50° for wind retrieval.

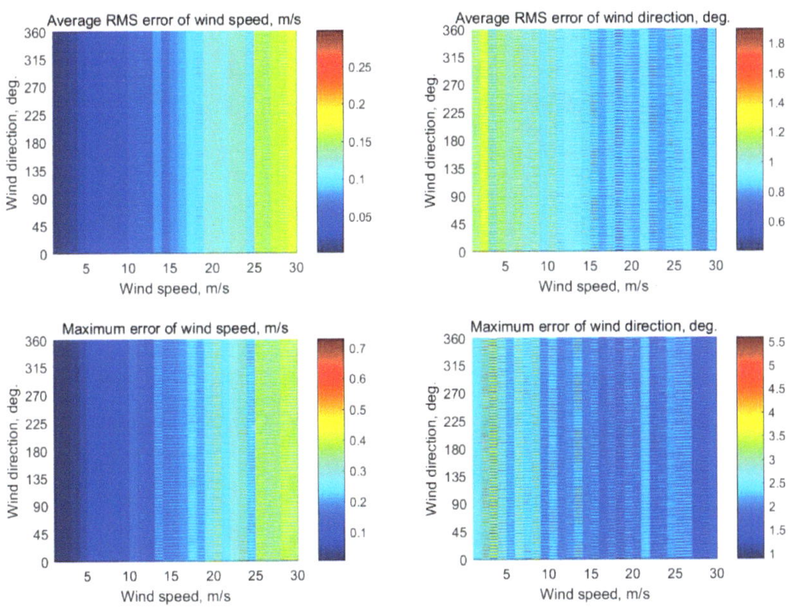


**Figure A6.** Simulation results of the use of only one incidence angle of 55° for wind retrieval.

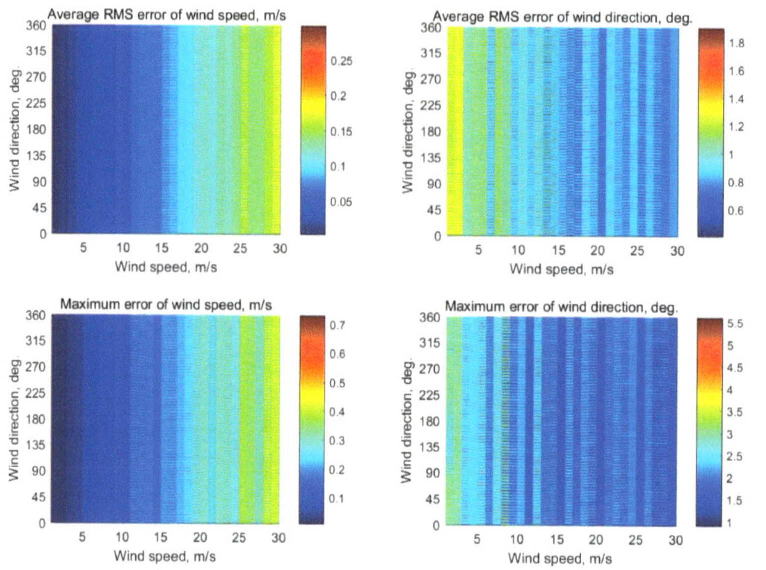


**Figure A7.** Simulation results of the use of only one incidence angle of 60° for wind retrieval.

## Appendix B

The simulation results of the wind retrieval with the system of Equation (4), when only two neighboring incidence angles are in use, are presented here. The results were obtained under the following conditions: The whole 360° azimuth circle observed was divided into $N = 72$ azimuth sectors (the azimuth sector width is 5°) in the directions of 0°, 5°, 10°, . . . , 355° relative to the aircraft flight direction assuming 0.2 dB instrumental noise and 87 integrated NRCS samples for each azimuth sector at wind speeds of 2–30 m/s. The results obtained for the incidence angle doublets of $\theta$ = (30°, 35°), $\theta$ = (35°, 40°), $\theta$ = (40°, 45°), $\theta$ = (45°, 50°), $\theta$ = (50°, 55°), and $\theta$ = (55°, 60°) are presented in Figures A8–A13, respectively.

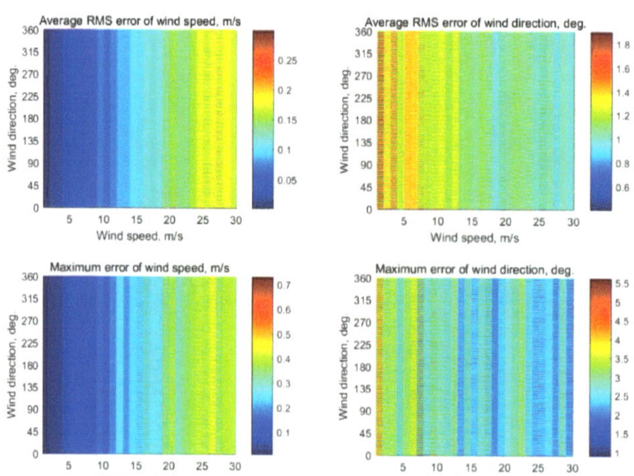


**Figure A8.** Simulation results of the use of two incidence angles of 30° and 35° for wind retrieval.

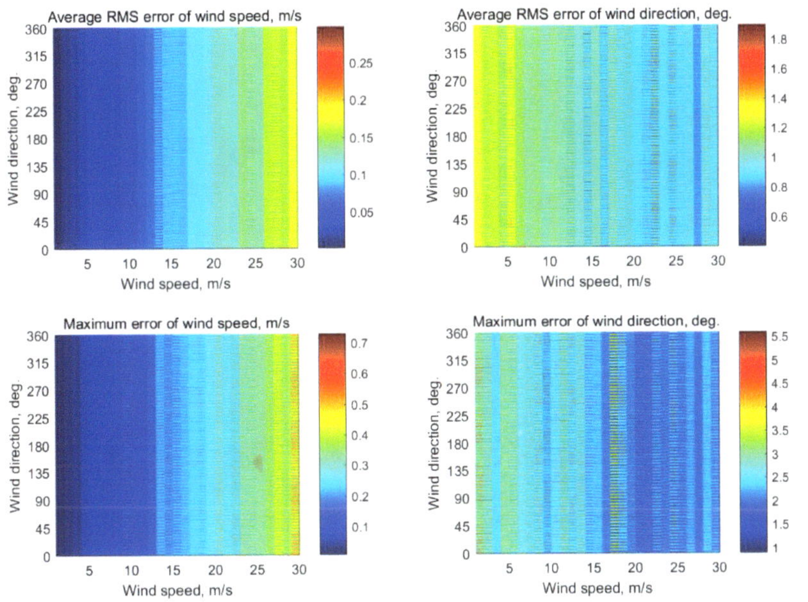


**Figure A9.** Simulation results of the use of two incidence angles of 35° and 40° for wind retrieval.

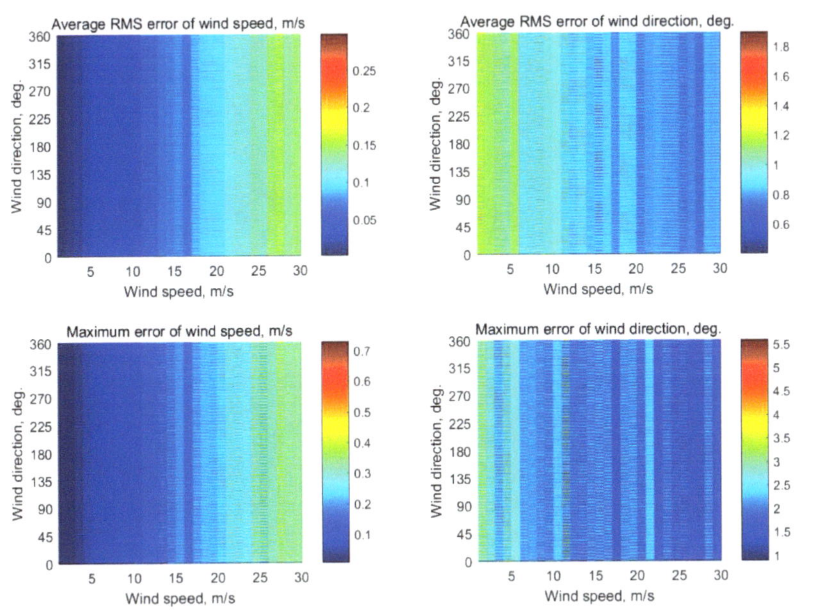


**Figure A10.** Simulation results of the use of two incidence angles of 40° and 45° for wind retrieval.

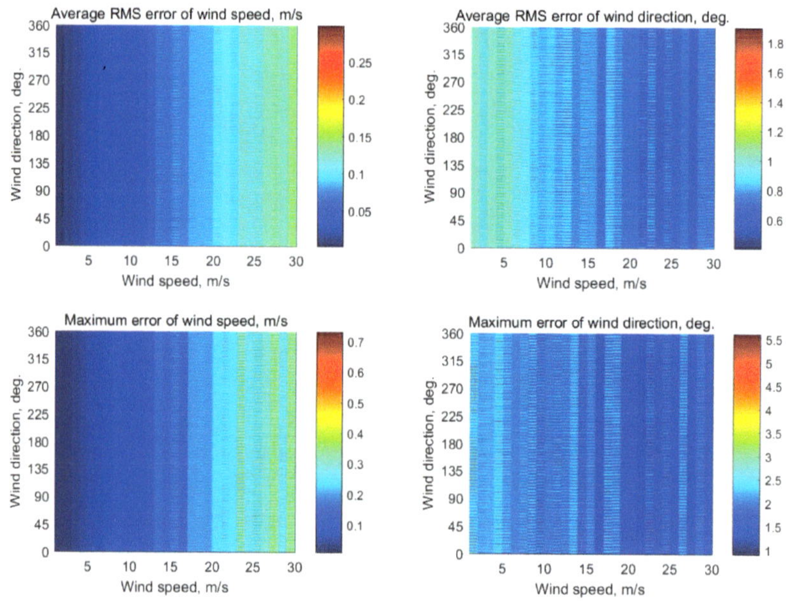


**Figure A11.** Simulation results of the use of two incidence angles of 45° and 50° for wind retrieval.

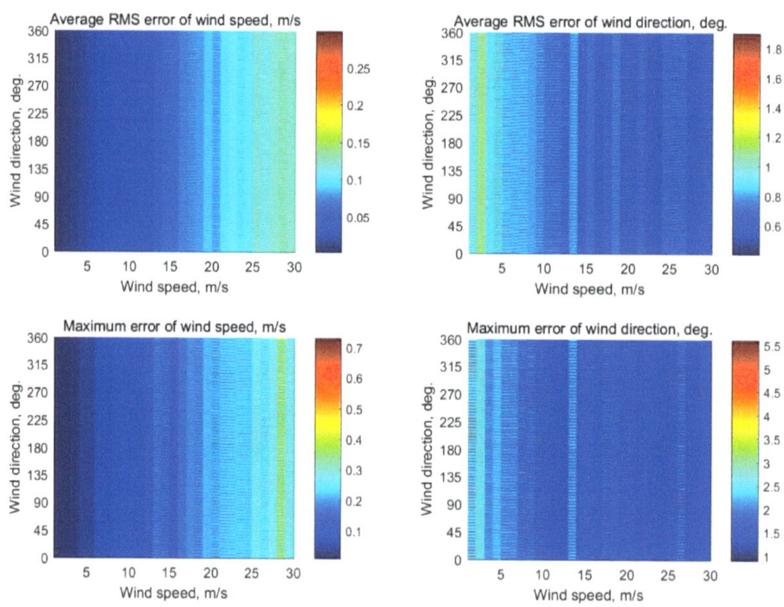


**Figure A12.** Simulation results of the use of two incidence angles of 50° and 55° for wind retrieval.

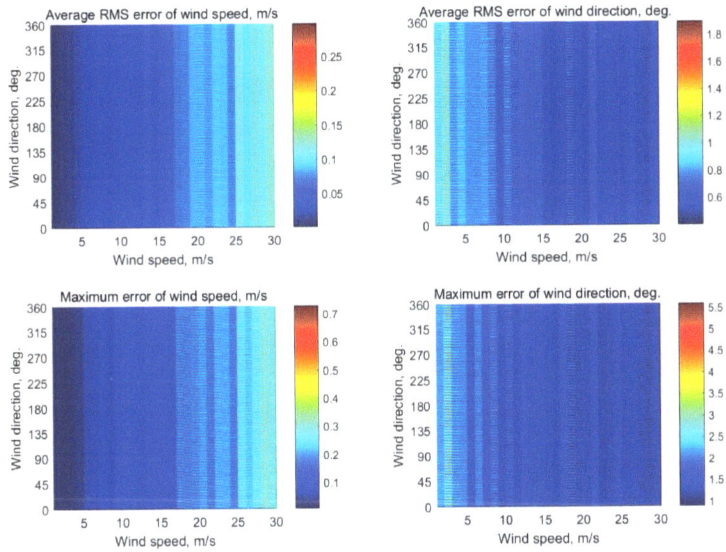


**Figure A13.** Simulation results of the use of two incidence angles of 55° and 60° for wind retrieval.

## Appendix C

The simulation results of the wind retrieval with the system of Equation (4), when only three neighboring incidence angles are in use, are presented here. The results were obtained under the following conditions. The whole 360° azimuth circle observed was divided into $N = 72$ azimuth sectors (the azimuth sector width is 5°) in the directions of 0°, 5°, 10°, ..., 355° relative to the aircraft flight direction assuming 0.2 dB instrumental noise and 87 integrated NRCS samples for each azimuth sector at wind speeds of 2–30 m/s. The results obtained for the incidence angle triplets of $\theta$ = (30°, 35°, 40°), $\theta$ = (35°, 40°, 45°), $\theta$ = (40°, 45°, 50°), $\theta$ = (45°, 50°, 55°), and $\theta$ = (50°, 55°, 60°) are presented in Figures A14–A18, respectively.

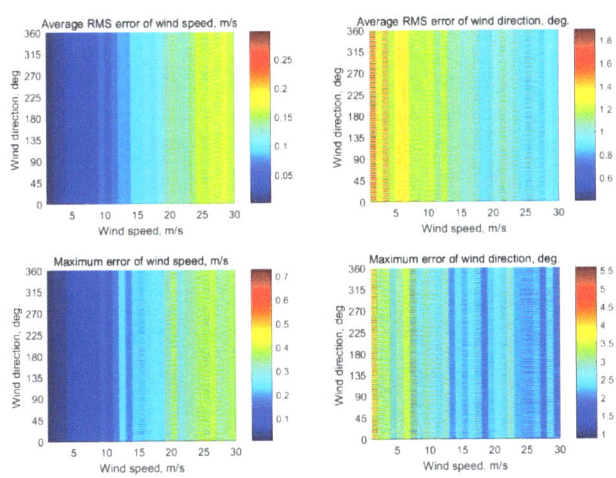


**Figure A14.** Simulation results of the use of three incidence angles of 30°, 35°, and 40° for wind retrieval.

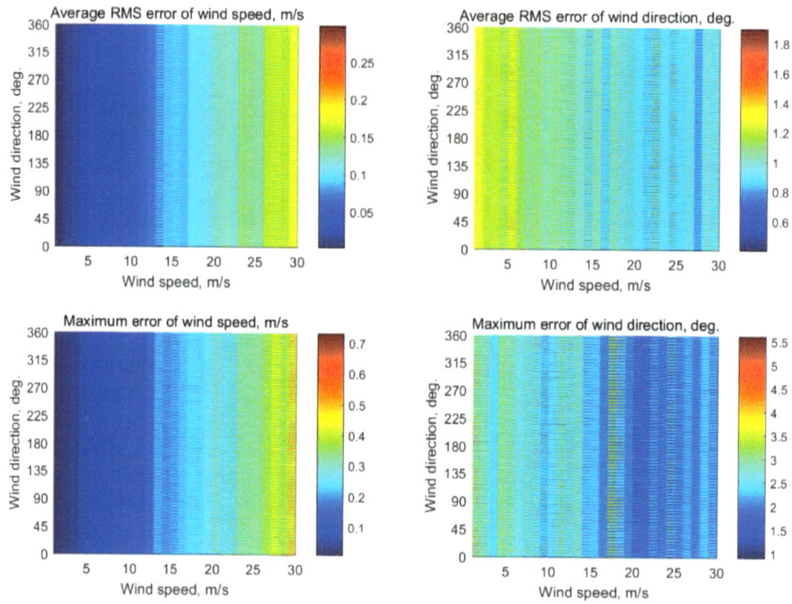


**Figure A15.** Simulation results of the use of three incidence angles of 35°, 40°, and 45° for wind retrieval.

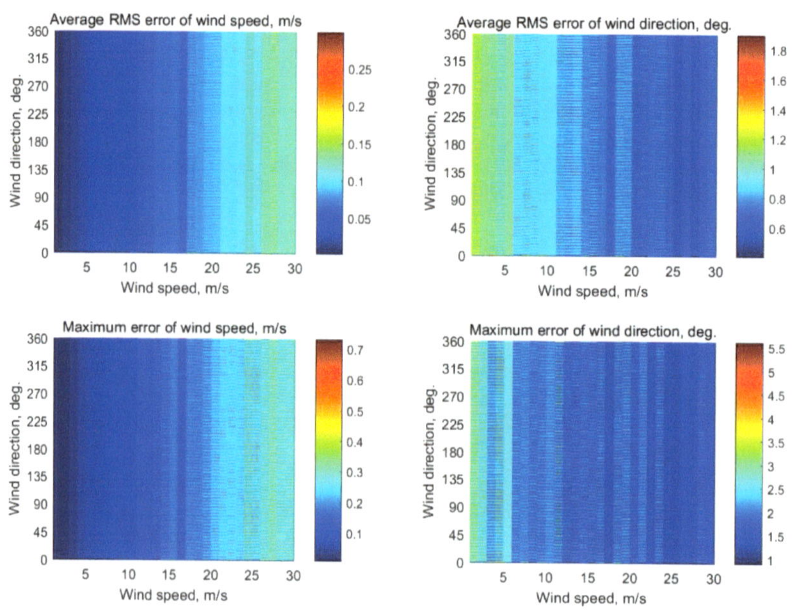


**Figure A16.** Simulation results of the use of three incidence angles of 40°, 45°, and 50° for wind retrieval.

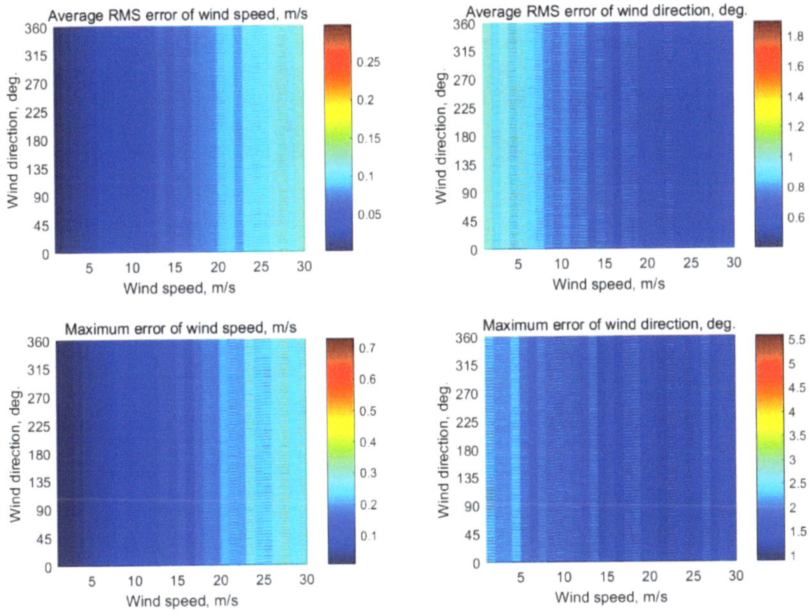


**Figure A17.** Simulation results of the use of three incidence angles of 45°, 50°, and 55° for wind retrieval.

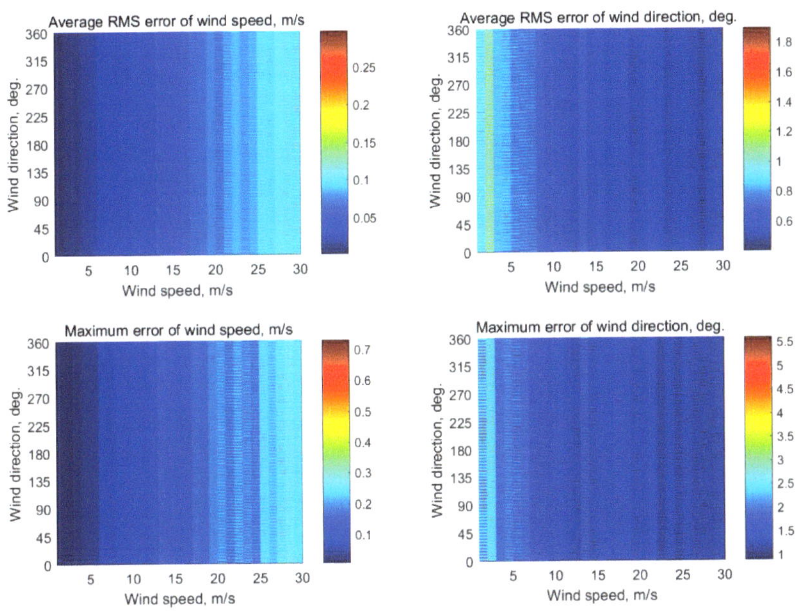


**Figure A18.** Simulation results of the use of three incidence angles of 50°, 55°, and 60° for wind retrieval.

## Appendix D

The simulation results of wind retrieval with the system of Equation (4), when only four neighboring incidence angles are in use, are presented here. The results were obtained under the following conditions: The whole 360° azimuth circle observed was divided into $N = 72$ azimuth sectors (the azimuth sector width is 5°) in the directions of 0°, 5°, 10°, ..., 355° relative to the aircraft flight direction assuming 0.2 dB instrumental noise and 87 integrated NRCS samples for each azimuth sector at wind speeds of 2–30 m/s. The results obtained for the incidence angle quartets of $\theta$ = (30°, 35°, 40°, 45°), $\theta$ = (35°, 40°, 45°, 50°), $\theta$ = (40°, 45°, 50°, 55°), and $\theta$ = (45°, 50°, 55°, 60°) are presented in Figures A19–A22, respectively.

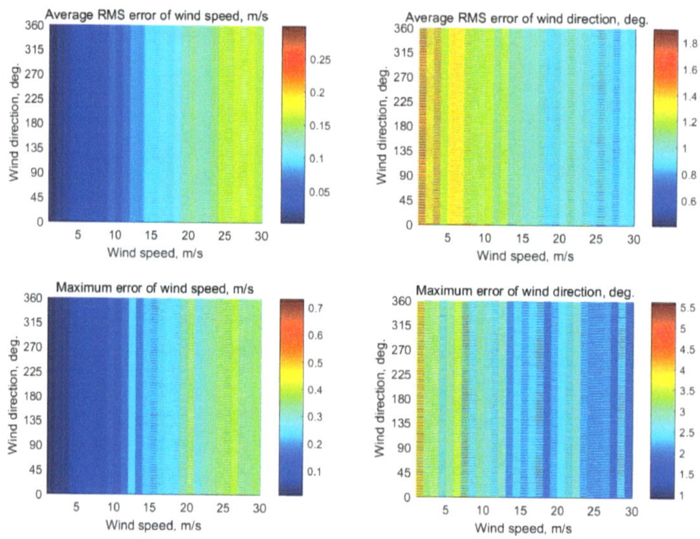


**Figure A19.** Simulation results of the use of four incidence angles of 30°, 35°, 40°, and 45° for wind retrieval.

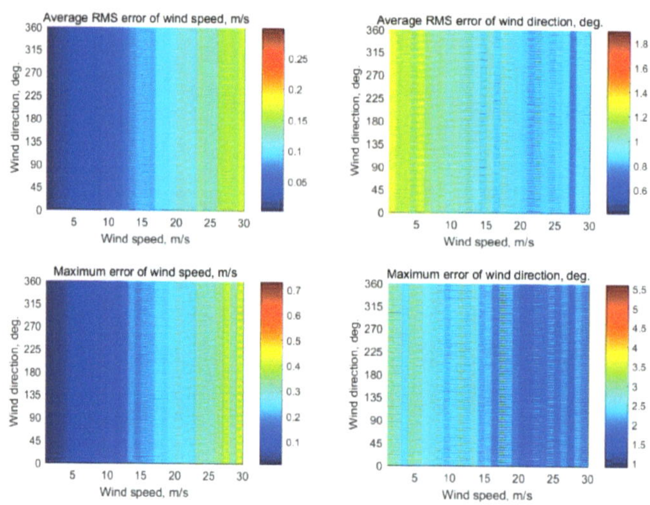


**Figure A20.** Simulation results of the use of four incidence angles of 35°, 40°, 45°, and 50° for wind retrieval.

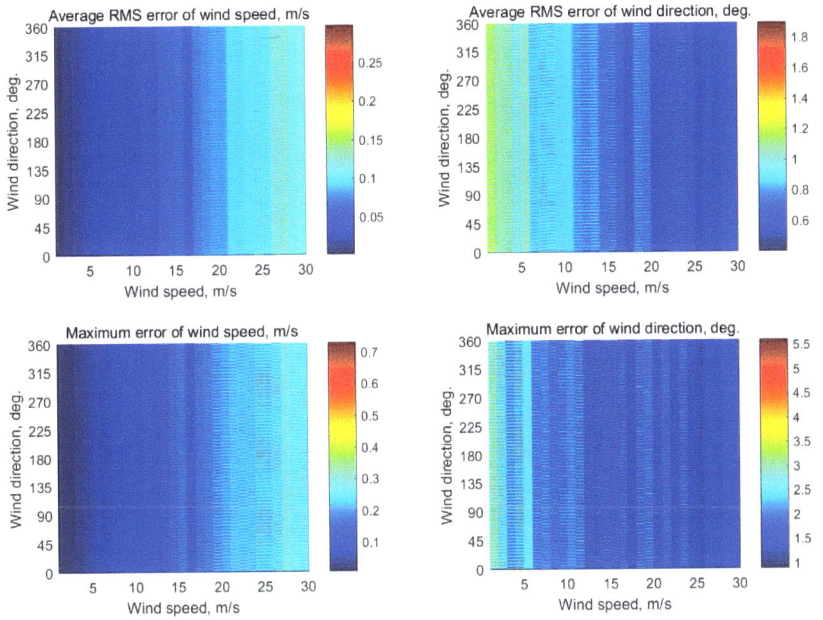


**Figure A21.** Simulation results of the use of four incidence angles of 40°, 45°, 50°, and 55° for wind retrieval.

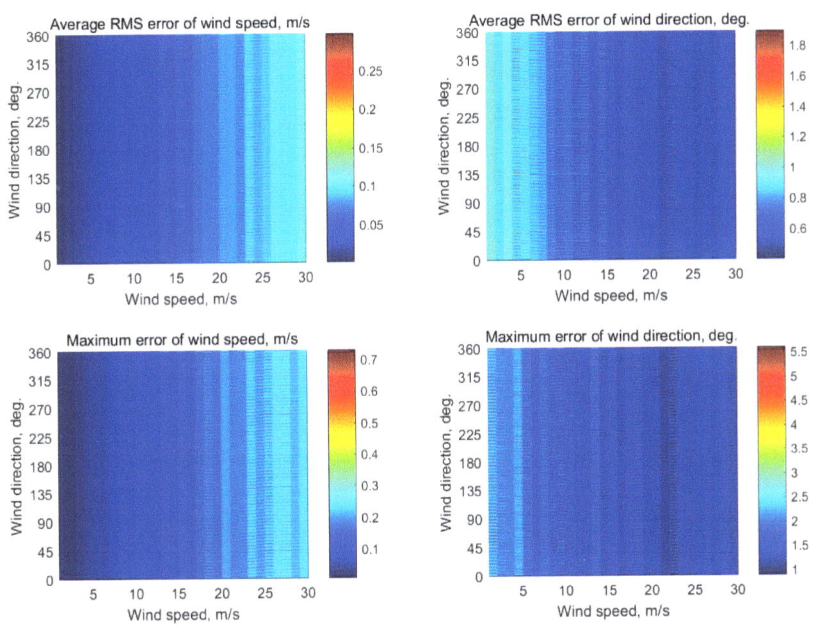


**Figure A22.** Simulation results of the use of four incidence angles of 45°, 50°, 55°, and 60° for wind retrieval.

## Appendix E

The simulation results of wind retrieval with the system of Equation (4), when seven neighboring incidence angles are in use, are presented here. The results were obtained under the following conditions: The whole 360° azimuth circle observed was divided into $N = 72$ azimuth sectors (the azimuth sector width is 5°) in the directions of 0°, 5°, 10°, . . . , 355° relative to the aircraft flight direction assuming 0.2 dB instrumental noise and 87 integrated NRCS samples for each azimuth sector at wind speeds of 2–30 m/s. The results obtained for the incidence angles of $\theta$ = (30°, 35°, 40°, 45°, 50°, 55°, 60°) are presented in Figure A23.

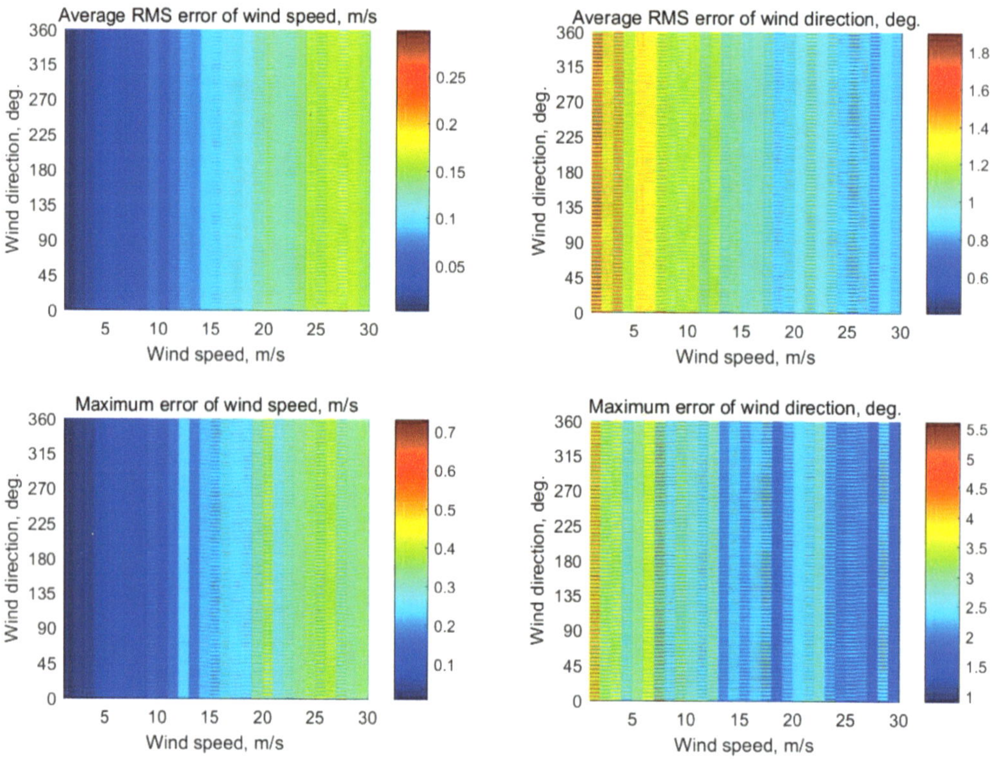


**Figure A23.** Simulation results of the use of seven incidence angles of 30°, 35°, 40°, 45°, 50°, 55°, and 60° for wind retrieval.

## Appendix F

The simulation results of wind retrieval with the system of Equation (4), when only three incidence angles are in use but with the highest incidence angle difference of 15° between the neighboring incidence angles (in the range of considered incidence angles of 30° to 60°), are presented here. The results were obtained under the following conditions: The whole 360° azimuth circle observed was divided into $N = 72$ azimuth sectors (the azimuth sector width is 5°) in the directions of 0°, 5°, 10°, . . . , 355° relative to the aircraft flight direction assuming 0.2 dB instrumental noise and 87 integrated NRCS samples for each azimuth sector at wind speeds of 2–30 m/s. The results obtained for the incidence angles of $\theta$ = (30°, 45°, 60°) are presented in Figure A24.

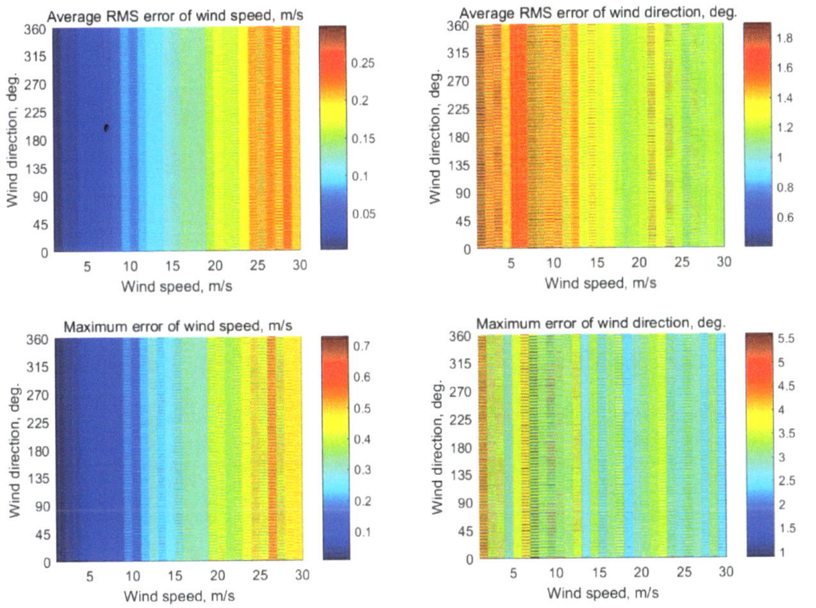


**Figure A24.** Simulation results of the use of three incidence angles of 30°, 45°, and 60° for wind retrieval.

## References

1. Hua, X.; Ono, Y.; Peng, L.; Cheng, Y.; Wang, H. Target detection within nonhomogeneous clutter via total Bregman divergence-based matrix information geometry detectors. *IEEE Trans. Signal Process.* **2021**, *69*, 4326–4340. [CrossRef]
2. Hurtado, M.; Nehorai, A. Polarimetric detection of targets in heavy inhomogeneous clutter. *IEEE Trans. Signal Process.* **2008**, *56*, 1349–1361. [CrossRef]
3. Petrov, N.; Le Chevalier, F.; Yarovoy, A.G. Detection of range migrating targets in compound-Gaussian clutter. *IEEE Trans. Aerosp. Electron. Syst.* **2018**, *54*, 37–50. [CrossRef]
4. Bounaceur, H.; Khenchaf, A.; Le Caillec, J.-M. Analysis of small sea-surface targets detection performance according to airborne radar parameters in abnormal weather environments. *Sensors* **2022**, *22*, 3263. [CrossRef]
5. Ward, K.D.; Tough, R.J.A.; Watts, S. *Sea Clutter: Scattering, the K Distribution and Radar Performance*; The Institution of Engineering and Technology: London, UK, 2006; p. 452. ISBN 978-1-84919-589-8.
6. Rodríguez, E.; Wineteer, A.; Perkovic-Martin, D.; Gál, T.; Stiles, B.W.; Niamsuwan, N.; Rodriguez Monje, R. Estimating ocean vector winds and currents using a Ka-band pencil-beam Doppler scatterometer. *Remote Sens.* **2018**, *10*, 576. [CrossRef]
7. Njoku, E.G. *Encyclopedia of Remote Sensing*; Springer: New York, NY, USA, 2014; p. 939. ISBN 978-0-387-36700-2.
8. Giovanangeli, J.-P.; Bliven, L.F.; Calve, O.L. A Wind-wave tank study of the azimuthal response of a Ka-band scatterometer. *IEEE Trans. Geosci. Remote Sens.* **1991**, *29*, 143–148. [CrossRef]
9. Snoeijl, P.; Van Halsema, D.; Oost, W.A.; Calkoen, C.; Jaehne, B.; Vogelzang, J. Microwave backscatter measurements made from the Dutch ocean research tower 'Noordwijk' compared with model predictions. In Proceedings of the IGARSS'92, Houston, TX, USA, 26–29 May 1992; pp. 696–698. [CrossRef]
10. Yurovsky, Y.Y.; Kudryavtsev, V.N.; Grodsky, S.A.; Chapron, B. Sea surface Ka-band Doppler measurements: Analysis and model development. *Remote Sens.* **2019**, *11*, 839. [CrossRef]
11. Wismann, V. Ocean windfield measurements with a rotating antenna airborne C-band scatterometer. In Proceedings of the 12th Canadian Symposium on Remote Sensing Geoscience and Remote Sensing Symposium, Vancouver, BC, Canada, 10–14 July 1989; pp. 1470–1473. [CrossRef]
12. Carswell, J.R.; Carson, S.C.; McIntosh, R.E.; Li, F.K.; Neumann, G.; McLaughlin, D.J.; Wilkerson, J.C.; Black, P.G.; Nghiem, S.V. Airborne scatterometers: Investigating ocean backscatter under low-and high-wind conditions. *Proc. IEEE* **1994**, *82*, 1835–1860. [CrossRef]
13. Jones, W.L. Early days of microwave scatterometry: RADSCAT to SASS. In Proceedings of the 2015 IEEE International Geoscience and Remote Sensing Symposium, Milan, Italy, 26–31 July 2015; pp. 4208–4211. [CrossRef]

14. Xu, X.; Dong, X.; Xie, Y. On-Board Wind Scatterometry. *Remote Sens.* **2020**, *12*, 1216. [CrossRef]
15. Nekrassov, A. Measurement of sea surface wind speed and its navigational direction from flying apparatus. In Proceedings of the MTS/IEEE Conference Oceans'97, Halifax, Nova Scotia, Canada, 6–9 October 1997; pp. 83–86. [CrossRef]
16. Karaev, V.Y.; Panfilova, M.A.; Titchenko, Y.A.; Meshkov, E.M.; Balandina, G.N.; Kuznetsov, Y.V.; Shlaferov, A.L. Retrieval of the near-surface wind velocity and direction: SCAT-3 orbit-borne scatterometer. *Radiophys. Quantum Electron.* **2016**, *59*, 259–269. [CrossRef]
17. Moore, R.K.; Fung, A.K. Radar determination of winds at sea. *Proc. IEEE* **1979**, *67*, 1504–1521. [CrossRef]
18. Masuko, H.; Okamoto, K.; Shimada, M.; Niwa, S. Measurement of microwave backscattering signatures of the ocean surface using X band and Ka band airborne scatterometers. *J. Geophys. Res. Ocean.* **1986**, *91*, 13065–13083. [CrossRef]
19. Nekrasov, A.; De Wit, J.J.M.; Hoogeboom, P. FM-CW millimeter wave demonstrator system as a sensor of the sea surface wind vector. *IEICE Electron.* **2004**, *1*, 137–143. [CrossRef]
20. Nekrasov, A. Airborne Doppler Navigation System Application for Measurement of the Water Surface Backscattering Signature. In Proceedings of the ISPRS TC VII Symposium—100 Years ISPRS, Vienna, Austria, 5–7 July 2010; The International Archives of the Photogrammetry, Remote Sensing and Spatial Information Sciences. Wagner, W., Székely, B., Eds.; 2010; Volume XXXVIII, Part 7A. pp. 163–168, Available online: https://www.isprs.org/proceedings/XXXVIII/part7/a/pdf/163_XXXVIII-part7A.pdf (accessed on 5 May 2022).
21. Tanelli, S.; Durden, S.L.; Im, E. Simultaneous measurements of Ku- and Ka-band sea surface cross sections by an airborne radar. *IEEE Geosci. Remote Sens. Lett.* **2006**, *3*, 359–363. [CrossRef]
22. Nekrasov, A.; Gamcová, M.; Kurdel, P.; Labun, J. On off-nadir wind retrieval over the sea surface using APR-2 or similar radar geometry. *Int. J. Remote Sens.* **2018**, *39*, 5934–5942. [CrossRef]
23. Nekrasov, A.; Khachaturian, A.; Veremyev, V.; Bogachev, M. Sea surface wind measurement by airborne weather radar scanning in a wide-size sector. *Atmosphere* **2016**, *7*, 72. [CrossRef]
24. Fernandez, D.E.; Kerr, E.; Castells, A.; Frasier, S.; Carswell, J.; Chang, P.S.; Black, P.; Marks, F. IWRAP: The Imaging Wind and Rain Airborne Profiler for remote sensing of the ocean and the atmospheric boundary layer within tropical cyclones. *IEEE Trans. Geosci. Remote Sens.* **2005**, *43*, 1775–1787. [CrossRef]
25. Coto, J.; Jones, W.L.; Heymsfield, G.M. Validation of the high-altitude wind and rain airborne profiler during the Tampa Bay rain experiment. *Climate* **2021**, *9*, 89. [CrossRef]
26. Hildebrand, P.H. Estimation of sea-surface wind using backscatter cross-section measurements from airborne research weather radar. *IEEE Trans. Geosci. Remote Sens.* **1994**, *32*, 110–117. [CrossRef]
27. Nekrasov, A.; Khachaturian, A. Towards the sea wind measurement with the airborne scatterometer having the rotating-beam antenna mounted over fuselage. *Remote Sens.* **2021**, *13*, 5165. [CrossRef]
28. Karaev, V.Y.; Kanevsky, M.B.; Balandina, G.N.; Meshkov, E.M.; Challenor, P.; Srokosz, M.; Gommenginger, C. A rotating knife-beam altimeter for wide-swath remote sensing of ocean: Wind and waves. *Sensors* **2006**, *6*, 620–642. [CrossRef]
29. Nekrasov, A.; Veremyev, V. Airborne weather radar concept for measuring water surface backscattering signature and sea wind at circular flight. *Nase More* **2016**, *63*, 278–282. [CrossRef]
30. Nekrasov, A.; Ouellette, J.D.; Majurec, N.; Johnson, J.T. A study of sea surface wind vector estimation from near nadiral cross-track-scanned backscatter data. *IEEE Geosci. Remote Sens. Lett.* **2013**, *10*, 1503–1506. [CrossRef]
31. Nekrasov, A.; Dell'Acqua, F. Airborne Weather Radar: A theoretical approach for water-surface backscattering and wind measurements. *IEEE Geosci. Remote Sens. Mag.* **2016**, *4*, 38–50. [CrossRef]
32. Nekrasov, A. Water-surface wind vector estimation by an airborne weather radar having a medium-size scanning sector. In Proceedings of the 14th International Radar Symposium IRS 2013, Dresden, Germany, 19–21 June 2013; Volume 2, pp. 1079–1084.
33. Unal, C.M.H.; Snoeij, P.; Swart, P.J.F. The polarization-dependent relation between radar backscatter from the ocean surface and surface wind vectors at frequencies between 1 and 18 GHz. *IEEE Trans. Geosci. Remote Sens.* **1991**, *29*, 621–626. [CrossRef]
34. Li, Z.; Stoffelen, A.; Verhoef, A. A generalized simulation capability for rotating-beam scatterometers. *Atmos. Meas. Technol.* **2019**, *12*, 3573–3594. [CrossRef]
35. Kramer, H.J. *Observation of the Earth and Its Environment: Survey of Missions and Sensors*, 4th ed.; Springer: Berlin/Heidelberg, Germany, 2002; p. 1509. [CrossRef]
36. Spencer, M.W.; Graf, J.E. The NASA scatterometer (NSCAT) mission. *Backscatter* **1997**, *8*, 18–24.
37. Nekrasov, A.; Khachaturian, A.; Abramov, E.; Popov, D.; Markelov, O.; Obukhovets, V.; Veremyev, V.; Bogachev, M. Optimization of airborne antenna geometry for ocean surface scatterometric measurements. *Remote Sens.* **2018**, *10*, 1501. [CrossRef]
38. Nekrassov, A. Sea surface wind vector measurement by airborne scatterometer having wide-beam antenna in horizontal plane. In Proceedings of the IGARSS'99, Hamburg, Germany, 28 June–2 July 1999; Volume 2, pp. 1001–1003. [CrossRef]
39. Nekrassov, A. On airborne measurement of the sea surface wind vector by a scatterometer (altimeter) with a nadir-looking wide-beam antenna. *IEEE Trans. Geosci. Remote Sens.* **2002**, *40*, 2111–2116. [CrossRef]
40. Nekrasov, A.; Khachaturian, A.; Veremyev, V.; Bogachev, M. Doppler navigation system with a non-stabilized antenna as a sea-surface wind sensor. *Sensors* **2017**, *17*, 1340. [CrossRef]
41. Nekrassov, A. Measurement of the sea surface wind vector by an airborne altimeter. In Proceedings of the IGARSS 2002, Toronto, Canada, 24–28 June 2002; pp. 1–3. [CrossRef]

42. Hans, P. *Auslegung und Analyse von Satellitengetragenen Mikrowellensensorsystemen zur Windfeldmessung (Scatterometer) Über Dem Meer und Vergleich der Meßverfahren in Zeit- und Frequenzebene. Von der Fakultät 2 Bauingenieur- und Vermessungswesen der Universität Stuttgart zur Erlangung der Würde eines Doktor-Ingenieurs Genehmigte Abhandlung;* Institut für Navigation der Universität Stuttgart: Stuttgart, Germany, 1987; 225 S. (In German)
43. Komen, G.J.; Cavaleri, L.; Donelan, M.; Hasselmann, K.; Hasselmann, S.; Janssen, P.A.E.M. *Dynamics and Modelling of Ocean Waves;* Cambridge University Press: Cambridge, UK, 1994; p. 532.

sensors

MDPI

Article

# Evaluation of a New Lightweight UAV-Borne Topo-Bathymetric LiDAR for Shallow Water Bathymetry and Object Detection

Dandi Wang [1,2], Shuai Xing [1,2,*], Yan He [3], Jiayong Yu [4], Qing Xu [1] and Pengcheng Li [1]

1 The Institute of Geospatial Information, Strategic Support Force Information Engineering University, 62 Science Road, Zhengzhou 450001, China; wdd_93@163.com (D.W.); 13937169139@139.com (Q.X.); lpclqq@163.com (P.L.)
2 Science and Technology on Near-Surface Detection Laboratory, 160 Tonghui Road, Wuxi 214035, China
3 Shanghai Institute of Optics and Fine Mechanics, Chinese Academy of Sciences, 390 Qinghe Road, Shanghai 201800, China; heyan@siom.ac.cn
4 The School of Civil Engineering, Anhui Jianzhu University, 292 Ziyun Road, Hefei 230601, China; yujiayongskd@163.com
* Correspondence: xing972403@163.com

**Abstract:** Airborne LiDAR bathymetry (ALB) has proven to be an effective technology for shallow water mapping. To collect data with a high point density, a lightweight dual-wavelength LiDAR system mounted on unmanned aerial vehicles (UAVs) was developed. This study presents and evaluates the system using the field data acquired from a flight test in Dazhou Island, China. In the precision and accuracy assessment, the local fitted planes extracted from the water surface points and the multibeam echosounder data are used as a reference for water surface and bottom measurements, respectively. For the bathymetric performance comparison, the study area is also measured with an ALB system installed on the manned aerial platform. The object detection capability of the system is examined with placed small cubes. Results show that the fitting precision of the water surface is 0.1227 m, and the absolute accuracy of the water bottom is 0.1268 m, both of which reach a decimeter level. Compared to the manned ALB system, the UAV-borne system provides higher resolution data with an average point density of 42 points/m$^2$ and maximum detectable depth of 1.7–1.9 Secchi depths. In the point cloud of the water bottom, the existence of a 1-m target cube and the rough shape of a 2-m target cube are clearly observed at a depth of 12 m. The system shows great potential for flexible shallow water mapping and underwater object detection with promising results.

**Keywords:** LiDAR; bathymetry; unmanned aerial vehicles; object detection; coastal mapping; full-waveform

## 1. Introduction

With the development of laser technology, airborne LiDAR bathymetry (ALB) [1] has shown great potential in shallow water surveys in recent decades. ALB as an active remote sensing technology can measure the two-way transmission time from the sensor to the water surface and bottom by emitting a green laser pulse (532 nm) to penetrate the water. Thus, ALB systems can obtain the positions of water surface and bottom simultaneously by equipped inertial navigation and global navigation satellite systems. Compared with the multibeam echosounder (MBES) technology used extensively in bathymetry, ALB mainly has two advantages in shallow water mapping. One advantage is that the airborne platform is not constrained by the underwater terrain and is suitable for shallow water and coastal mapping [2]. The other advantage is that the swath width is determined by the flight altitude and scan angle rather than the water depth, so it can still maintain high efficiency in shallow water mapping [3]. Optical mapping sensors (RGB, multispectral cameras) also can be used for bathymetric surveying [4,5]. The application of Structure from Motion (SfM) technology [6] makes it possible to generate a 3D point cloud from overlapped aerial images, greatly improving the accuracy and spatial resolution of the bathymetric mapping. Although optical mapping sensors have advantages in low cost and high spectral resolution compared to ALB systems, it is difficult to retrieve water depths when the water bottom texture is unclear or invisible, and the depth accuracy is sensitive to the water surface conditions and water clarity [7].

Because the green laser requires sufficient pulse energy to penetrate the water, most ALB systems are designed for manned platforms, such as CZMIL, HawkEye 4X, and VQ-880-G. For eye safety considerations, the beam divergence of ALB systems is larger than topographic LiDAR and is normally greater than 1 mrad [8]. The flight altitude of the manned platform is commonly kept in the range of 300–500 m, so at a typical scan angle of 15°, a beam divergence of 1 mrad corresponds to a footprint diameter of 0.3–0.5 m [9]. The large beam divergence results in a large surface footprint size, which will influence the geometric precision.

Mounting the ALB system on unmanned aerial vehicles (UAVs) is an effective solution for collecting high-resolution measurements in some special cases, such as small island and reef mapping, underwater object detection, flood management, etc. The flying height of multi-copter UAVs can be set in the range of 20–50 m, reducing the footprint diameter to at least one-tenth that of manned aerial platforms. In addition, compared to manned aircraft, UAVs have the advantages of lower cost, easier operation, and freer airspace, allowing for more flexible and agile measurements [10,11]. However, due to the decrease in flight altitude, the swath width of UAVs is narrow (around 20 m), so they are only suitable for small areas survey.

In recent years, some ALB systems can be mounted on UAVs, called UAV-borne ALB systems. According to the sensor weight, the existing UAV-borne ALB systems can be divided into two categories. One is the lightweight system with a weight of about 5 kg, such as *RIEGL* BathyDepthFinder-1 (BDF-1) [12] and *ASTRALite* edge™ [13]. The lightweight systems can be mounted on standard UAV platforms like the *DJI* Martice 600 Pro [14]. The benefit of the system is that it can operate at low altitudes with foot point diameters down to the centimeter level. However, laser energy is restricted due to the weight limitation, resulting in a weak bathymetric capability. Furthermore, the normal ALB systems commonly use circular or elliptical scanning patterns, while the lightweight systems use different ones. For example, the BDF-1 operates at a fixed 15° off-nadir, so it can only generate profile data [15]. Edge™ uses a rectilinear scanning pattern resulting in a wide range of incident angles of the laser beam, which may affect the stability of the received signals [16]. The other type of UAV-borne ALB system is the compact system that weighs more than 10 kg and can be mounted on large-scale UAVs, such as *Fugro* RAMMS [17] and *RIEGL* VQ840-G [18]. The compact systems are miniaturized complete systems with comparable bathymetric performances and high point density. The maximum measured water depth can reach 3 Secchi depths (SD) [19], and the point density is approximately 20–50 points/$m^2$ [20].

The existing UAV-borne ALB systems have been successfully applied to map inland rivers and lakes [20–22], showing the potential to offer better accuracy and higher spatial resolution data. In 2021, a new experimental lightweight UAV-borne ALB system, Mapper4000U, was developed by the Shanghai Institute of Optics and Fine Mechanics (SIOM), Chinese Academy of Science, and Strategic Support Force Information Engineering University. The weight of the sensor system is about 5 kg, so that it can be mounted on *DJI* Matrice 600 Pro. The system can simultaneously emit near-infrared (NIR, 1064 nm) and green (532 nm) laser pulses with an elliptical scanning pattern, ensuring efficient measurements and stable data acquisition. Equipped with the dual-wavelength laser, it is capable of classifying land and water signals with high accuracy [23,24], which is crucial for areas with blurred or irregular land-water boundaries. Another advantage of the dual-wavelength laser is that it can improve the accuracy of the water surface position when the water is turbid [25,26]. Recently, a flight test was conducted with the Mapper4000U in Dazhou Island, China. In this paper, the experimental results were presented, and compared to the MBES and manned ALB data acquired during the same period. The performances of the Mapper4000U, including the capability of bathymetry and object detection, were thoroughly evaluated.

The main contribution of this paper is providing a comparative analysis of the bathymetric performance of a new lightweight UAV-borne ALB system using the field data (UAV-borne ALB data, manned ALB data and MBES data) in a coastal area. The local plane fitting errors of the water surface points are analyzed to assess the relative accuracy of the water surface measurements of the UAV-borne ALB system. To estimate the bathymetric accuracy of the system, the water bottom measurements obtained by the system and MBES are directly compared based on the ellipsoid heights. The system is also compared with a manned ALB system in bathymetric performances, including the survey efficiency, point spatial distribution, measured depth range, and small object detection. Furthermore, the underwater object detection capability of the system is analyzed by manually placing ideal target cubes.

The rest of the paper is organized as follows. In Section 2, a detailed description of the Mapper 4000U is present. The study area, field datasets, and the methods for data processing and evaluation are introduced in Section 3. The experimental results are presented in Section 4 and discussed in Section 5. Finally, Section 6 summarizes this paper.

## 2. Mapper4000U

The Mapper4000U is a lightweight and compact topo-bathymetric LiDAR system and is also a miniaturization of the *SIOM* Mapper5000 [27], which is designed for manned aerial platforms. The appearance of the sensor system is shown in Figure 1. There are two options to power it: using an external battery or a drone power supply. A solid-state drive (SSD) can be easily inserted to save data and unplugged to read data. The Mapper4000U can be connected to a digital camera for trigger control and can be integrated with a position and orientation system (POS) to build a comprehensive ALB system. As the sensor system only weighs 4.4 kg, it is flexible in selecting the POS and power supply solution to meet the payload restrictions of UAV platforms and the measurement requirements. The detailed system parameters are listed in Table 1.

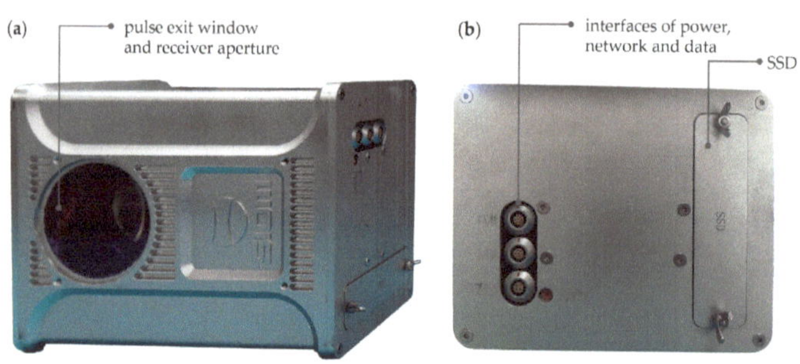


**Figure 1.** The appearance of the Mapper4000U; (**a**) the bottom of the system with the pulse exit window and receiver aperture for the near-infrared (NIR) and green laser; (**b**) the front side of the system with the interfaces and solid-state drive (SSD) socket.

**Table 1.** Mapper4000U specifications.

| Pulse Repetition Frequency | Pulse Energy | Scan Rate | Size | Weight |
|---|---|---|---|---|
| 4 kHz | 12 μJ@1064 nm<br>24 μJ@532 nm | 15 lines/s | 235 mm × 184 mm × 148 mm | 4.4 kg |

The laser can emit both NIR and green pulses at a frequency of 4 kHz. A rotating scanner is used to create an elliptical scanning pattern with a scan angle of ±15° along-track and about ±12° cross-track. The beam divergence of the green laser is reduced from 1 mrad to 0.5 mrad because of the lower pulse energy. As a result, the device has a small footprint diameter (5 cm at an altitude of 50 m), ensuring that small features of the underwater terrain can be captured.

The receiver has two receiving channels, including an avalanche photodiode (APD) channel for the NIR laser pulse and a photomultiplier tube channel (PMT) for the green laser pulse. The PMT has a wide effective response range to avoid problems of signal saturation in extremely shallow waters [28] and ensure the ability of above 1.5 SD penetration. All the channels are sampled at a rate of 1G samples/s, i.e., the waveform data are recorded at an interval of 1 ns, corresponding to a slant distance of about 0.15 m in air and 0.11 m in water.

Compared to the other lightweight UAV-borne ALB systems, the Mapper4000U has better performances with a dual-wavelength laser and a rotating scanner. At the same time, the Mapper4000U still maintains the weight of a lightweight system and is more flexible than the existing compact UAV-borne ALB systems. In this paper, the UAV-borne ALB system is first presented and tested in a coastal area, and the bathymetric performance and object detection capability are evaluated.

## 3. Materials and Methods

### 3.1. Study Area

The study area Dazhou Island is located in the southeast of Wanning City, Hainan Province, China. The island is about 5 km off the coast of Wanning and consists of two ridges, a small northern ridge, and a large southern ridge, with a 500 m long beach in the middle. The study area is exactly in the sea between the two ridges. The water is clear, and the SD is in the range of 5–10 m. The underwater terrain is a continuous and gentle slope, and the water bottom is mostly covered by sand, which is suitable for assessing the bathymetric performance of the system.

An 8-day survey for Dazhou Island was conducted from 25 September 2021 to 2 October 2021, as shown in Figure 2. The dates of data acquisition are shown in Table 2. The study was designed for a comprehensive performance evaluation of Mapper4000U, including a precision analysis, an accuracy assessment based on MBES reference data, a direct comparison between the manned (Mapper5000) and unmanned (Mapper4000U) ALB, and an object detection test. The field data contains two flight strips of the Mapper4000U, referred to as S1 and S2, one flight strip of the Mappper5000, and the MBES measurements.

### 3.2. Field Data

In this experiment, the Mapper4000U was powered by a separate battery and equipped with a POS (*NovAtel* SPAN-IGM-S1). The total weight of the system, including the sensor, POS, SSD, and battery, was 5.6 kg. The system was mounted on the *DJI* Matrice 600 Pro, as shown in Figure 3a,b. Using the standard batteries, a flight can last 10–15 min with approximately 30% power left. The flight altitude was kept at 50 m, and the swath width on the water surface of each strip was 21 m. The width of the overlap between S1 and S2 was 4 m, so the range of the measurement area was about 38 m × 1200 m. The specific data parameters are shown in Table 3.

**Table 2.** Data acquisition dates.

| Mapper4000U | Mapper5000 | MBES | Target Placement |
|---|---|---|---|
| 26 September | 2 October | 29 September | 25 September |

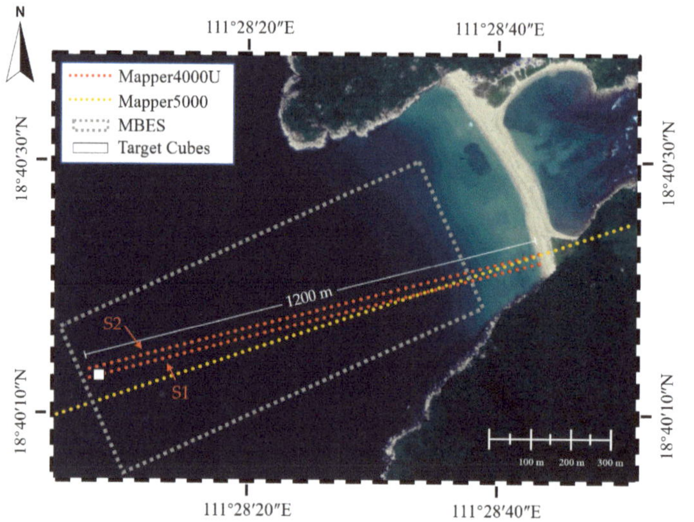


**Figure 2.** Study area. The red dotted lines denote the flight strips of the Mapper4000U, the yellow dotted line denotes the flight strip of the Mappper5000, the range of the multibeam echosounder (MBES) measurements is in the gray dashed box, and the location of the target cubes is denoted by the white box.

**Figure 3.** Photos of the Mapper4000U data acquisition and target placement; (**a**) Mapper4000U mounted on DJI Matrice 600 Pro; (**b**) detailed installation positions of the payloads; (**c**) a photo of the 2-m fabric cube; (**d**) an aerial image of the target cubes.

**Table 3.** Data acquisition parameters of the Mapper4000U and Mapper5000.

| System | Altitude | Speed | Swath Width | Point Density | Flight Duration [1] |
|---|---|---|---|---|---|
| Mapper4000U | 50 m | 5 m/s | 21 m | 42 points/m$^2$ | 225 s |
| Mapper5000 | 375 m | 65 m/s | 201 m | 0.38 points/m$^2$ | 22 s |

[1] Only the flight time of each strip was counted.

In addition to the UAV survey, the experiment was carried out in the following steps:

1. For the accuracy assessment, the study area was also measured by a MBES (*Hydro-tech Marine* MS400), and a digital bathymetric model (DBM) [29] with high-resolution (0.2 m) was generated using the supporting software. The DBM was used as reference data of water bottom points in this experiment. The geographic coordinates of both the reference DBM and Mapper4000U survey points used the WGS84 ellipsoid and were projected into UTM zone 49 N.
2. For the bathymetric performance comparison, a Mapper5000 survey were performed a few days after the UAV survey. The flight altitude, speed, and swath width are much higher than that of the UAV, but the point density is sharply decreased. The data acquisition parameters are compared in Table 3. In data processing, a depth-adaptive waveform decomposition method was used for signal detection, and a post-processing software developed by the manufacturer was used for point cloud generation, including geo-calibration and refraction correction [27]. For comparison, the measured points were also transformed to the geographic coordinates (WGS84).
3. Small object detection capability. Two fabric targets, a 1-m white cube and a 2-m white cube, were placed in water one day before the UAV survey, and the location of the targets was measured at the same time. The cubes gradually sank to a depth of about 12 m, which was deeper than the Secchi depth.

### 3.3. Data Processing of the UAV-Borne ALB

The data processing of the Mapper4000U mainly consists of four steps:

1. POS data processing. The observations from the POS mounted on the platform and a temporary reference station were processed using Waypoint Inertial Explorer8.8 Software to estimate the flight trajectory.
2. Waveform data processing. The full waveforms were sampled and recorded by the receiver. To extract the signals, a fast and simplified processing method was applied to the received waveforms (see below for a detailed description).
3. System calibration. The system calibration was conducted in a nearby village. Six strips of Mapper4000U data were collected, and a number of control points were measured by RTK GNSS survey. Thus, the extrinsic error was corrected based on the planar calibration model [30].
4. Coordinates calculation. Based on the flight trajectory, the extrinsic parameters, and the refraction correction model [27], the detected signals were converted to the 3D point cloud in the WGS_1984_UTM_Zone_49N coordinate.

We simplified and modified the wave decomposition method proposed in [27] to fit the new system and improve the efficiency. Waveforms collected by Mapper4000U were classified into three categories, as shown in Figure 4, and then independently processed.

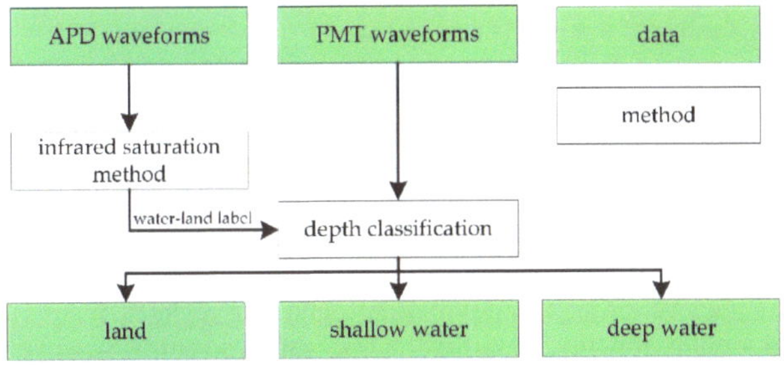


**Figure 4.** Workflow of the waveform classification. APD: avalanche photodiode; PMT: photomultiplier tube.

Based on the infrared saturation method [31], waveforms were first classified into two groups, land and water, using APD waveforms. In the infrared saturation method, the saturation time of the APD waveforms is defined as the duration beyond the maximum effective value (saturated value) of the receiver. The saturation time $t_{SAT}$ can be expressed as follows:

$$t_{SAT} = \sum_{t=0}^{N} w_{APD}(t) \geq W_{SAT}, \quad (1)$$

where $w_{APD}$ is the APD waveform and $W_{SAT}$ denote the upper bound of the effective response range. The $W_{SAT}$ is the saturation value of the receiver's output voltage, which can be easily determined from a waveform with a saturated signal. If $t_{SAT}$ is greater than or equal to the saturation threshold, the waveform will be labeled as "land"; otherwise, it will be labeled as "water" (see Figure 5a). The saturation threshold was set to 4 ns in this paper.

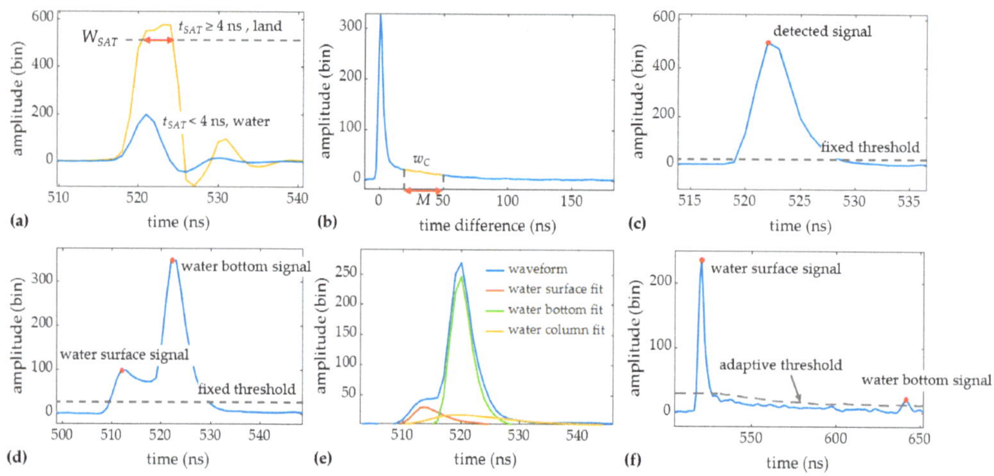


**Figure 5.** Schematic illustration of the waveform data processing; (**a**) water-land classification based on the infrared saturation method; (**b**) depth classification for water waveforms using the truncated water column scattering waveform (i.e., $w_C$); (**c**) signal detection of the land waveform using the fixed threshold; (**d**) signal detection of the shallow water waveform using the fixed threshold; (**e**) empirical decomposition of the shallow water waveform; (**f**) signal detection of the deep-water waveform using the adaptive threshold.

For "water" waveforms, further classification is needed because of the different characteristics of shallow and deep water waveforms. In the depth classification method [27], the similarity between the received waveforms and the water column scattering is measured, as shown in Figure 5b. The similarity $S$ can be calculated as:

$$S = \min\{R(t)\}, \quad (2)$$

$$R(t) = \frac{1}{M}\sum_{m=1}^{M} [w_C(m\tau) - w_{PMT}(m\tau + t)], \quad (3)$$

where $w_{PMT}$ denotes the PMT waveform, $w_C$ is the truncated water column scattering waveform extracted from $w_{PMT}$, $M$ is the length of $w_C$, and $\tau$ is the sampling interval. As the shape of the water column scattering changes little in a small survey area, the deep-water waveform will have a high similarity, while the shallow water waveform, where the water column scattering is mainly covered by the surface and bottom returns, will have a low similarity. Based on the value of $S$, the "water" waveforms were classified into two parts, "shallow water" and "deep water".

PMT waveforms in the three categories were processed as follows:

- Land waveforms processing. Considering the gentle coastal terrain, only the moving-average algorithm [32] and a signal detection method with a fixed maximum threshold [33] are performed on the land waveforms, as shown in Figure 5c.
- Shallow water waveforms processing. Shallow water waveforms are first processed using the same methods as the land waveforms for signal detection. If the number of detected signals is greater than or equal to 2, the first signal will be recorded as the water surface signal, and the last signal will be recorded as the water bottom signal, as shown in Figure 5d. If only one signal can be detected, which occurs when the water depth is extremely shallow, the waveform will be decomposed based on an empirical model [27] to extract the water surface and bottom signal, as shown in Figure 5e.
- Deep-water waveforms processing. Denoising is the key to deep water waveforms processing, while the existing waveform filtering methods cannot appropriately deal with the high-intensity noise in the water column scattering. Thus, the fixed threshold in the signal detection method is replaced by a depth-adaptive threshold derived from the truncated water column scattering waveform [27], which greatly reduces the effect of noise in the water column scattering. The intensities of the detected signal are subtracted from the depth-adaptive threshold, and the two signals with the highest strength are selected as the water surface and bottom signal, as shown in Figure 5f.

### 3.4. Methods for the Evaluation

The performance of the Mapper4000U were evaluated in four aspects, and the evaluation methods in [20] were partially used here.

1. Precision: Analysis of the relative accuracy of the measurements. The water surface points of each strip were searched in 1 m × 1 m grids based on the planimetric coordinates. In each grid, the points were fitted to a plane, and the distance from the point to the plane was calculated to estimate the ranging precision. In addition, the DBMs of the water bottom were generated via the moving least squares interpolation and compared in the overlapping area of the two strips to evaluate the consistency of the data.
2. Accuracy: Assessment of the UAV system's bathymetric accuracy. As the bathymetric LiDAR and MBES only measure instantaneous depths, the depth measurements cannot be directly compared. Therefore, the accuracy was evaluated by comparing the ellipsoid heights of the measured water bottom points with the reference values derived from the DBM generated by MBES measurements in the same planimetric coordinates.
3. Bathymetric performance: A comparison of the Mapper4000U and Mapper5000 for bathymetry (including point density, maximum depth penetration, and object detection capability). The point distribution and average density were both considered, and the profiles of the water bottom point clouds obtained from the UAV-borne system and manned platform system were compared. The maximum detected depths of the systems were estimated with the Secchi depth as reference. The capability of small object detection was examined in shallow and deep water.
4. Object detection capability: The target cube points were extracted from the water bottom point cloud of Mapper4000U and were fitted and projected to planes. To assess the detection accuracy, the distances from the points to the fitted plane and the shape of the projected points were statistically analyzed.

## 4. Results

### 4.1. Precision

The precision of the measured water surface reflects the ranging accuracy of the system and affects the refraction correction results, which indirectly influences the accuracy of the water bottom points. Since the water surface height is instantaneous, the reference data should be measured simultaneously. Although the APD channel can obtain a more accurate water surface, the instability of the signals makes it impossible to generate a high-density reference point cloud [34]. Based on the spatial coherence of the water surface [35], we obtained the references of the water surface height by local plane fitting. The error of the water surface height (i.e., $\delta_S$) is estimated according to the height difference, which is defined as:

$$\delta_S = h_S - h'_S, \tag{4}$$

where $h_S$ is the ellipsoid height of the measured water surface points and $h'_S$ is the ellipsoid height of the fitted plane in the same planimetric coordinate.

The histograms of the height deviations of the two strips, S1 and S2, are shown in Figure 6. The error distributions of S1 and S2 are approximately the same, which follow the Gaussian distribution and are centered around the zero value. In contrast, the distribution of S1 is slightly concentrated.

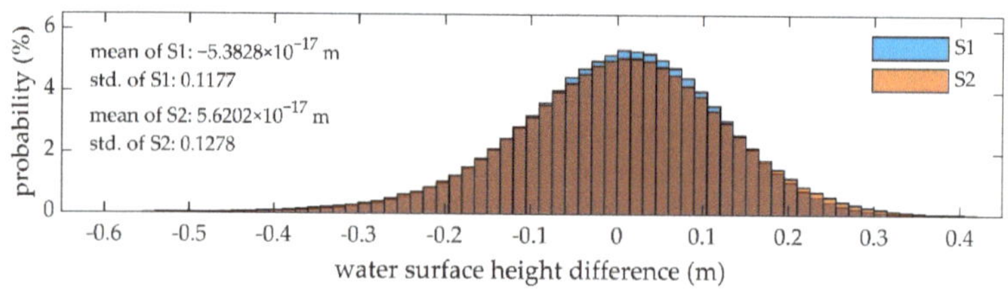


**Figure 6.** Probability distributions of the water surface height differences between measured points and fitted planes of S1 (blue) and S2 (orange); std.: standard deviation.

The errors of the water surface height are presented numerically in Table 4. The deviation between the mean water surface heights of S1 and S2 is 5.5 mm. The mean water level elevation (MWLE) in the study area at the time of the UAV survey is about −7.9478 m. The root mean square error (RMSE) of S1 is slightly lower than that of S2. The overall RMSE of the water surface height is below 0.13 m and 98% of the height errors are within ±0.3 m.

**Table 4.** Statistics of water surface ellipsoid height including mean, RMSE, and percentage of the error within ±0.3 m.

| Strip | Mean of Height [m] | RMSE [m] | $\mid \delta_S \mid < 0.3$ m [%] |
|---|---|---|---|
| S1 | −7.9452 | 0.1177 | 98.49 |
| S2 | −7.9507 | 0.1278 | 97.49 |
| Sum | −7.9478 | 0.1227 | 98.01 |

To assess the consistency of the measurements, the height differences of the water bottom points in the overlapping area of S1 and S2 were analyzed. First, the water bottom points of the two strips were converted from the UTM coordinate to a relative coordinate system with the $x$-axis parallel to the direction of the strips. Then two digital elevation models with 1 m resolution for S1 and S2 (DBM1 and DBM2) were created along-track using moving least squares interpolation, respectively. The DBM1 was subtracted from the DBM2 within the overlapping range to obtain the height differences of the water bottom (dDBM).

Figure 7 shows the water bottom measurements of S1 and S2 with the location of the overlapping area and DBM1, DBM2, and dDBM in the overlapping range. By comparing DBM1 and DBM2, it can be observed that the elevations of the water bottom points in the same planimetric coordinates are almost equal, showing high consistency between the measurement results of the two strips. In dDBM, the height deviations are mainly within ±0.2 m. The height differences are regionally distributed, with negative values concentrated in the upper left and positive values in the middle and rightmost.

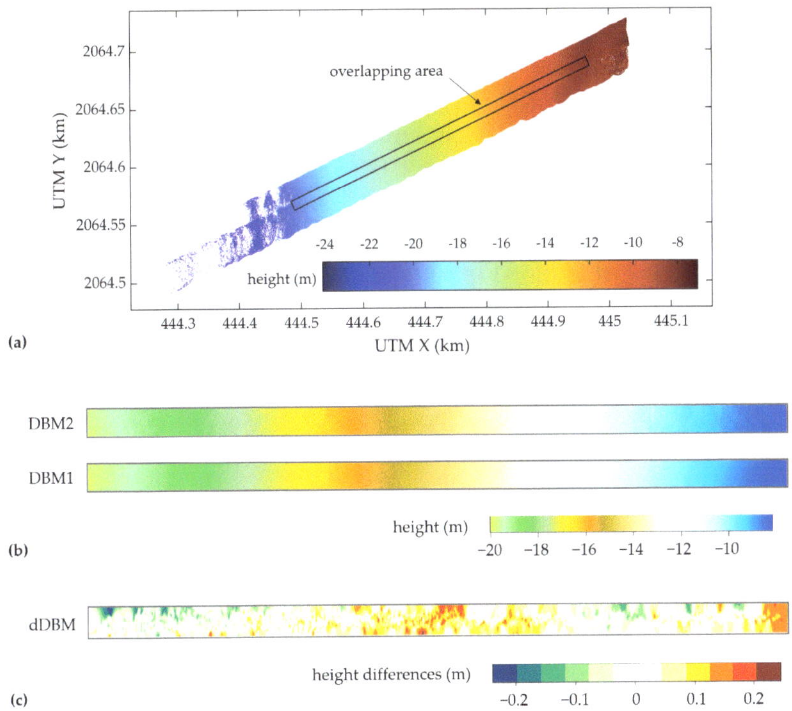


**Figure 7.** Comparison of water bottom height in the overlapping area; (**a**) color-coded map of the water bottom height of S1 and S2 with the overlapping area marked in the black box; (**b**) DBM1 and DBM2 in the overlapping area; (**c**) the height differences between the two strips in the overlapping area; DBM: digital bathymetric model.

In Figure 8, a profile of the dDBM in the vertical direction is plotted. The DBM1 is averagely 2.72 cm higher than the DBM2, and the dispersion of the height difference is 7.11 cm. According to the DBM1 and DBM2 shown in Figure 7, the water depth decreases gradually from left to right, while the height differences of the water bottom rise first, then fall, and finally rise again. Thus, the trend of the height differences is not consistent with that of the water depths.

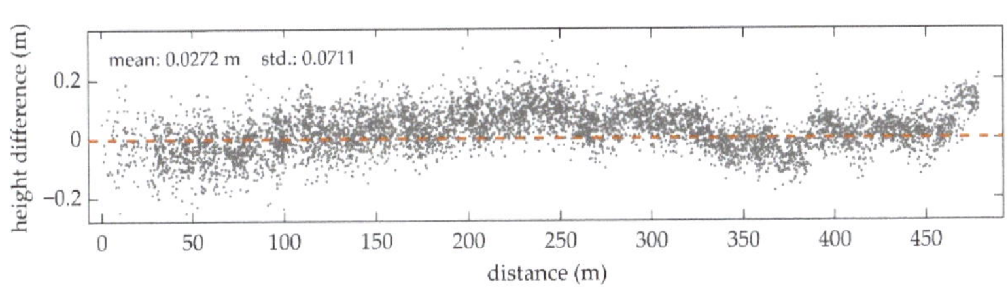


**Figure 8.** Distribution of the water bottom height differences along the direction of the strips, where the horizontal coordinates are the distances from the grids to the leftmost grid in the strip direction, and the vertical coordinates are the elevation value of dDBM.

### 4.2. Accuracy

The accuracy of the measured water bottom ellipsoid heights $h_B$ directly represents the bathymetric accuracy of the system. In this experiment, the ellipsoid height references

$h'_B$ were obtained from the DBM of MBES. The error of the water bottom height (i.e., $\delta_B$) can be expressed as

$$\delta_B = h_B - h'_B. \quad (5)$$

Figure 9 shows the spatial distribution of $\delta_B$, where the right side is the water near the beach (i.e., the water depth gradually increases from right to left). The majority of the errors are negative, indicating that the ellipsoid height of the water bottom point is generally lower than the reference value. The water bottom points with large errors are mainly distributed in deep waters, especially in S2, where some water bottom points with the $\delta_B$ close to ±0.5 m exist. Comparing the error distributions of S1 and S2, it can be seen that the accuracy of S1 is higher, which is consistent with the precision assessment of the water surface height in Section 4.1.

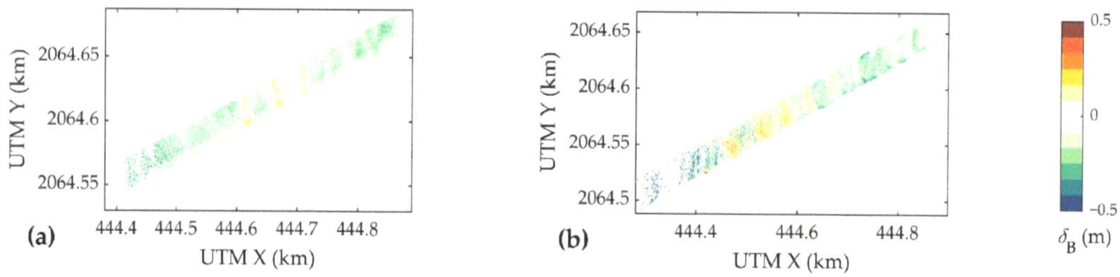


**Figure 9.** Color-coded maps of the water bottom height errors of (**a**) S1 and (**b**) S2.

Table 5 summarizes the statistical analysis of the water bottom height. According to the range of the height and the MWLE (−7.9478 m) estimated in Section 4.1, the range of the detectable depth of Mapper4000U is 0–16 m. The SD in the survey area was around 8.33 m which was estimated by the aerial images simultaneously acquired by *DJI* Phantom 4 pro. Based on the position of the deepest underwater object that could be visually observed from the images, the SD in the survey area was approximated by the corresponding water depth acquired from the manned ALB system. Thus, the maximum detectable depth corresponds to 1.7–1.9 SD. The overall RMSE of the water bottom height is 0.1268 m, which is larger than that of the water surface height. Different from the water surface points, the accuracy of the water bottom points in S1 is distinct from that in S2, with a difference of 3.45 cm. The mean of $\delta_B$ for both S1 and S2 is negative, resulting in an overestimation of depth which is also found in the measurements of VQ840-G [20]. For the requirement of the vertical accuracy (within ±0.3 m), more than 96% of the water bottom points are qualified.

**Table 5.** Statistics of water bottom ellipsoid height including maximum (max.), minimum (min.), RMSE, mean of the error and percentage of the error within ± 0.3 m.

| Strip | Max. of Height [m] | Min. of Height [m] | RMSE [m] | Mean of $\delta_B$ [m] | $\mid \delta_B \mid < 0.3$ m [%] |
|---|---|---|---|---|---|
| S1 | −7.9500 | −22.1639 | 0.1075 | −0.0520 | 98.48 |
| S2 | −8.0665 | −24.1038 | 0.1420 | −0.0705 | 93.89 |
| Sum | −7.9500 | −24.1038 | 0.1268 | −0.0615 | 96.11 |

The profiles of the point clouds of S1 and S2 along the center of the strip are shown in Figure 10a,b, respectively, where all the target points are plotted as water bottom on the profile of S2 regardless of the distances to the profile. Most of the water bottom points are below the references, which is also observed in Figure 9 and Table 5. When the height is below −20 m, the error of the water bottom point increases significantly. The dispersion of the water bottom points in S2 is greater, which is consistent with the statistical results in Table 5. Both in S1 and S2, the land points and water bottom points at the land-water interface are well connected, and the topography is continuous with slight height differences. The depth of the targets is about 12 m, which exceeds the SD in the survey area.

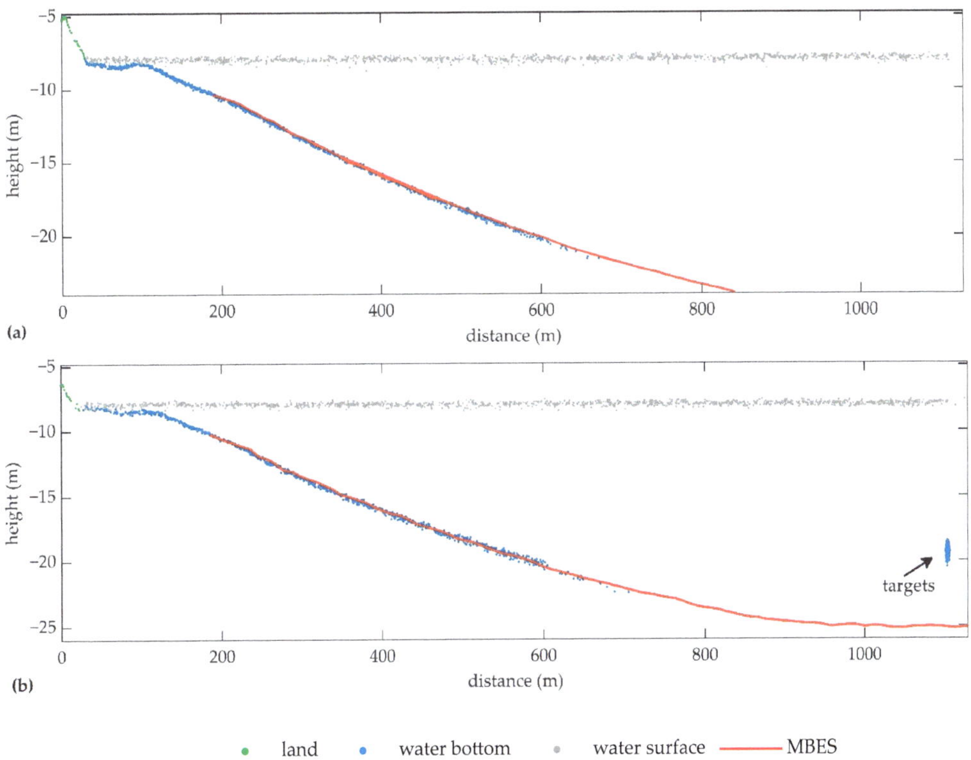


**Figure 10.** Profiles of the 3D point clouds of (**a**) S1 and (**b**) S2 colored by classification, and the DBM obtained by MBES, where the horizontal coordinates are the distances to the point with maximum X-coordinate in the UTM coordinate system.

### *4.3. Bathymetric Performance*

The bathymetric performances of the UAV-borne ALB system (Mapper4000U) and the manned ALB system (Mapper5000) were compared in this experiment. Figure 11 presents the distributions of water surface points, where the red dots denote the point cloud of S1, and the blue dots are the point cloud of the selected strip of Mapper5000 (for location, see Figure 2). Owing to the elliptical scanning pattern, the point clouds of both Mapper4000U and Mapper5000 are unevenly distributed, with points sparse in the middle and dense on the sides. There are two scanning lines at one position, and the trajectories of the forward and backward scanning are crossed. The point density of Mapper4000U is significantly higher than that of Mapper5000, while the swath is narrower. According to the specifications shown in Table 3, the average point density of Mapper4000U is 42 points/m$^2$, which is 110 times that of Mapper5000, but the swath is only 21 m, which is only one-tenth that of Mapper5000.

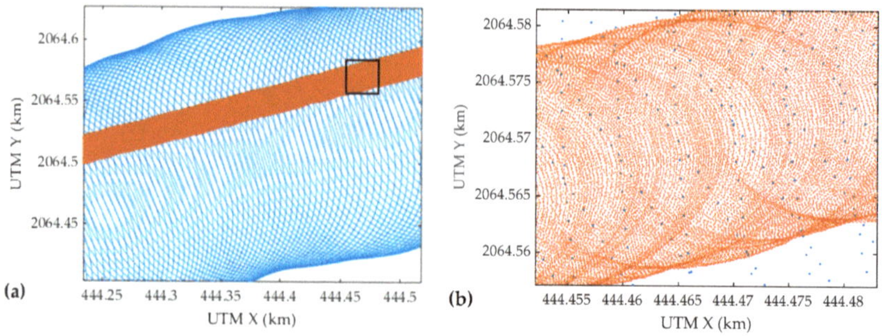


**Figure 11.** Distributions of water surface points acquired by the Mapper4000U (red dots) and the Mapper5000 (blue dots) on the XY plane; (**a**) the overall distribution; (**b**) the detailed distribution (zoom in the black box).

Figure 12 shows the cross-section of the point clouds of S2 and Mapper5000 strip along the track of S2. Both systems can acquire land, water surface, and water bottom points simultaneously. The MWLE varies slightly due to the different dates of data acquisition. The heights of the land and water bottom points obtained from the two platforms are very close. Most of the water bottom points obtained from the UAV are above those obtained from the manned aircraft. Since the overestimation of water depth in UAV data was found in Section 4.2, this problem is still present and even more serious in the manned platform data. Based on the heights of the water surface points, the MWLE at the time of the Mapper5000 survey can be estimated, and the result is −8.36 m. Since the minimum height of the water bottom points of Mapper5000 in the study area is −34.33 m, the maximum detectable depth of the system is 25.97 m, corresponding to 3 SD.

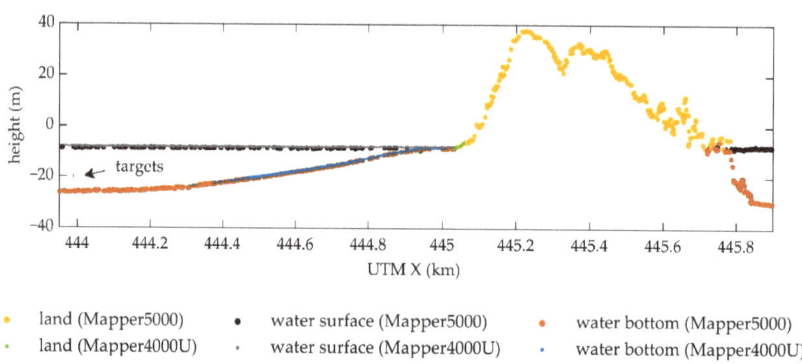


**Figure 12.** Profiles of the 3D point clouds obtained by the Mapper4000U and Mapper5000.

Although the ALB system mounted on the manned platform can measure deeper waters compared to the UAV system, it is unable to keep the detailed topographic features due to the low point density and large footprint size. For example, a circular mound and the placed targets can be clearly identified in the point cloud of the Mapper4000U. The shape of the circular mound is reserved, while the targets suffer a severe shape deformation. However, the circular mound and the placed targets are failed to be found in the point cloud of the Mapper5000, as shown in Figure 13.

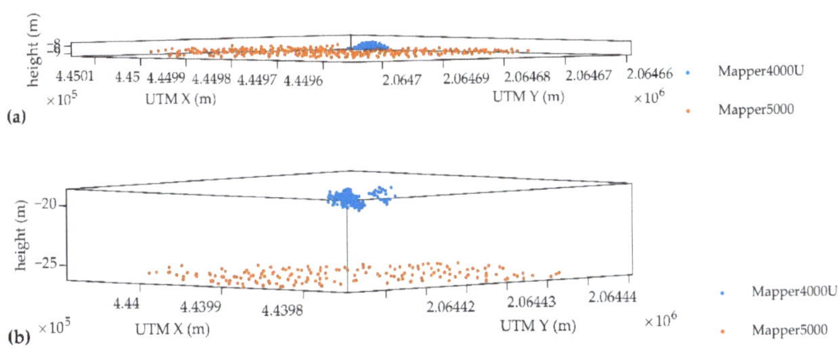


**Figure 13.** Perspective view of 3D point clouds of (**a**) a 3 m width circular mound and (**b**) the placed targets.

## 4.4. Object Detection Capability

In the point cloud of S2, two clusters of target points were detected, and their corresponding targets were judged based on the volumes of the clustered points. Two planes can be clearly observed in the point cloud of the 2-m cube, so the points are fitted separately, namely P1 and P2. Since the shape of the point cloud of the 1-m cube is hard to be identified, all the points are fitted to a plane, namely P3.

The plane fitting results of the target points are presented in Figure 14. For the convenience of presentation, the 3D fitting results are plotted under a relative coordinate system, i.e., the origin of the UTM is translated on the XY plane, and 2D fitting results are plotted under the relative coordinate system established for each fitted plane individually. For the 2-m target cube, the angle between the normal vectors of P1 and P2 is 36°, the number of points in each plane is approximately 80, and the fitting error is less than 0.17 m. From the projection of the points in the fitted plane, we can see that the coverage area of the points is larger than the standard 2-m square, and the deformation is obvious. From the profile, the error of the target points is within ±0.5 m, but the target points are more divergent than the water bottom points. For the 1-m target cube, the number of the point cloud is only 25. From the projection in the fitted plane, the shape of the points is similar to a combination of two 1-m squares, and the area is close to 1 m × 2 m. However, from the direction perpendicular to the plane, it is not easily identified as a composition of two faces. If the point clouds on the left and right sides are fitted with lines, respectively, the angle between the lines is about 115°.

(a)

(b)

(c)

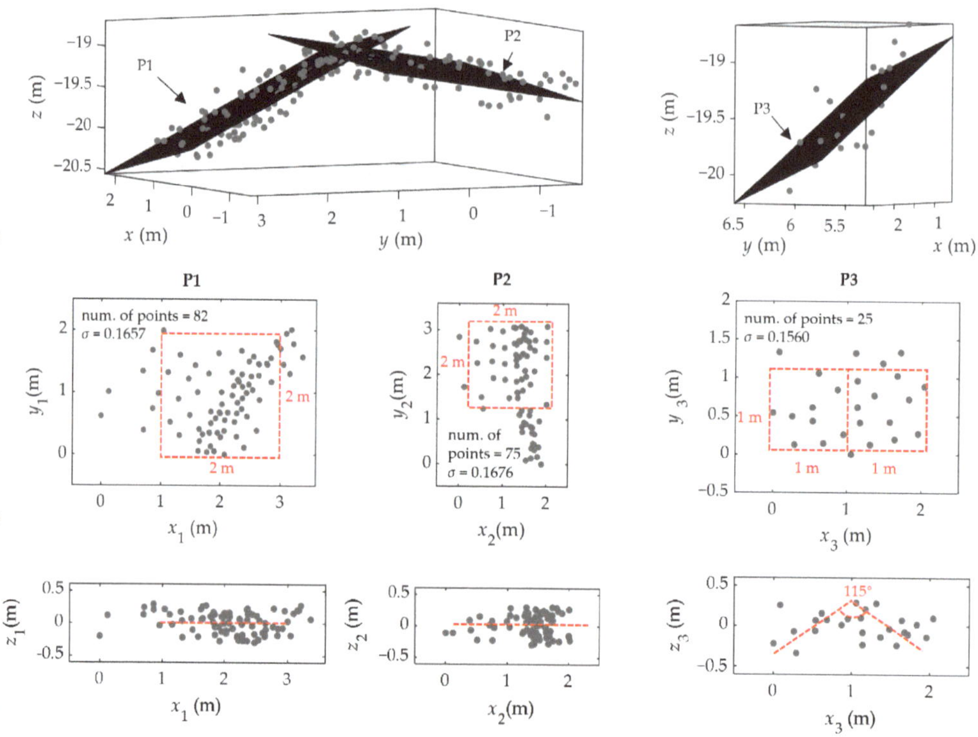


**Figure 14.** Results of the target points fitting; (**a**) perspective view of 3D point clouds of the 2-m target cube and 1-m target cube with the fitted planes P1–P3; (**b**) target points projected to P1–P3 with the dispersion of the point-to-plane distance (i.e., $\sigma$); (**c**) vertical section of the target points.

## 5. Discussion

### *5.1. Environmental Effects on Water Surface Detection*

For the precision of the water surface, the error distributions of S1 and S2 are similar, as shown in Figure 6, indicating that the error is independent of the measurement location and flight direction. In addition to the systematic errors, the influence of environmental factors cannot be ignored. For example, when the water surface signal is weak, the detected peak is a mixed peak of the water surface signal and the water column backscattering, so the position of the water surface signal cannot be accurately estimated, which is called the water surface uncertainty of the green laser [1,26]. As a result, the detected water surface will be lower than the actual water surface. As the slope and roughness of the water surface are random, the intensity of the water surface signal cannot be predicted, introducing a random error in the water surface signal detection. Furthermore, the sampling rate of the receiver is 1 GHz (c.f. Table 1); that is, the interval between two adjacent samples in the waveform is 1 ns, which corresponds to a distance of 0.15 m on the slant path in the air. Since the water surface signal position acquired by the signal detection method can only be an integer, an approximation error is introduced to the water surface, reaching a theoretical maximum of 0.075 m. Although waveform decomposition can eliminate the effect of the water surface uncertainty and obtain accurate results, it is quite time-consuming and is not suitable for emergency usage. One of the feasible solutions is to increase the sampling rate of the receiver, and the other is to optimize the current processing method to fast estimate the accurate position of the water surface signal.

### 5.2. Consistency between the Adjent Strips

Because of the narrow swath of the UAV-borne system, the matching error of the strips needs to be considered. In the evaluation of the strip consistency, the height difference of the water bottom between the two unmatched strips is generally within ±0.2 m. Therefore, the results can be directly used for DBM generation without strip adjustment if the accuracy of the products is not strictly required. Regional distribution of the height difference is observed in Figure 8, showing that there may be a small deviation between the two strips. It is expected to produce a more accurate DBM by correcting this error through the strip adjustment methods such as the iterative closest point (ICP) algorithm.

### 5.3. Impact of In-Water Path Calculation on Water Bottom

In ALB, the water bottom positions are derived from the water surface measurements and the in-water path of the laser pulse. Unlike the error distribution of the water surface points, the error distribution of the water bottom points is not uniform, and the error rises as the water depth increases. The calculation of the in-water path consists of three elements, the starting point (i.e., the water surface point), the path direction, and the propagation velocity. The accuracy of the water surface measurements has been discussed in Section 5.1. The water surface, as the interface between air and water, will affect the refraction correction. If the detected water bottom signal is accurate, a low water surface position induced by water surface uncertainty will have a downward vertical shift impact on the water bottom point, which may be responsible for the overestimation of water depths (c.f. Table 5).

However, the random error caused by water surface uncertainty cannot explain the problem that the water bottom error grows with the water depth. The error induced by the refraction direction accumulates with depth, which may be the main reason for the large errors in deep waters. Waves can affect the slope of the water surface and thus change the direction of the in-water path, which severely influences the UAV-borne ALB system with a small footprint size [36,37]. In [35], a 3D water surface profile model was built to predict the water surface slope, which is expected to solve this problem.

Another factor that affects the in-water path is the propagation velocity. The refractive index of the water determines the in-water propagation direction and velocity of the laser pulse and is usually fixed at 1.33. Recent studies have found that the fixed refractive index of 1.33 is inappropriate [20]. Using a group velocity instead of the fixed refractive index can reduce the range-dependent bias of the water depth [38]. Therefore, the calculation of the in-water path has a significant effect on water bottom, especially in deep waters, which need to be carefully corrected.

### 5.4. Bathymetric Performance Comparison

From the examined bathymetric performances of the manned and UAV-borne ALB system, it is found that manned ALB systems have the advantages of fast flight speed, extensive coverage, high efficiency, strong power energy, and a wide range of detectable depth. The drawbacks are the low point density and large footprint size, making it difficult to measure the complex features of the terrain accurately. Furthermore, the placed targets are invisible in the point cloud of the Mapper5000. One possibility is that the interval between the measurement and target placement lasts a few days (c.f. Table 2), and the target shifted during this period. Another possibility is that the points are too sparse, and the target points are mistakenly removed as noise during the point filtering process. According to the average point density, a 2-m square corresponds to about 1.5 points. If the targets do not locate at the edge of the strip, where the point density is relatively high, they will be difficult to detect. The advantages of the UAV-borne ALB system are the extremely high point density and small footprint size, which ensure sufficient detected points and accurate measurements, besides the maximum detectable depth of 2 SD can meet the demands of most shallow water surveys. The circular mound in shallow water and the placed targets in deep water all can be clearly observed in the point cloud (see Figure 13). In addition, the UAV platform is easy to operate and can finish the measurement of two 21 m × 1200 m strips in one flight. The flight duration is only 10 min including take-off, landing, and calibration. For a nearshore area with a range of 500 m × 1200 m, at least 30 flight strips need to be measured with 20% strip overlap, which takes 150 min in total. Although the UAV platform is less efficient than the manned aerial platform according to the speed and swath width, UAVs, especially multi-copter UAVs, which can vertical take-off and landing, have a few requirements for environmental conditions. As a result, the flight preparation time is greatly reduced, and the measurement efficiency is indirectly improved.

The bathymetric performances of the UAV-borne ALB system and MBES can also be compared in Figure 10. The advantages of MBES are high accuracy, high point density, and the ability to measure extremely deep waters. However, the minimum detectable depth of MBES is limited by the navigable area of the ship-borne platform. ALB can obtain land, water surface, and water bottom point clouds simultaneously. With the equipped NIR laser, the land and water points can be precisely distinguished. The excellent shallow-water measurement capability of ALB can complement the data of MBES. For example, in this experiment, the minimum depth measured by MBES is 2.31 m, while the maximum depth measured by the UAV-borne ALB system is 16 m. There is enough overlapping area that allows one to merge the measurements of MBES and ALB.

In summary, ALB is good at shallow water surveys, where the manned ALB is suitable for regular large area survey tasks, and the UAV-borne ALB is applicable for detailed surveys in small areas, besides MBES can help ALB to complete measurements in deep water. However, both ALB and MBES lack spectral information, limiting their further applications. The combination of ALB, MBES, and optical mapping sensors has been extensively researched and widely applied for inland water surveying [39–41] and coastal shallow water mapping [42,43]. Considering that the lightweight UAV-borne ALB system allows flexible installation, the integration of the system and optical cameras can be mounted on various UAV platforms. The fusion of the high spatial resolution LiDAR point cloud and high spectral resolution images in future work will provide a promising solution to coastal management.

### 5.5. Evaluation of the Object Detection Capability

To assess the object detection capability, two fabric cubes were placed in the water. Although both the 1-m cube and 2-m cube are identified in the point cloud, deformations and irregular shapes are also obvious. Two faces can be observed in the 2-m cube, but the angle between them is not equal to 90°. For the 1-m cube, it is impossible to identify the original shape from the point cloud. However, the underwater topography can be clearly observed in shallow water, such as a 3 m width circular mound (see Figure 13). Therefore, the accumulation errors induced by the in-water path are possibly responsible for the deformations of the target cubes. Because the bathymetric accuracy of the system gradually reduces with the increase in water depth (c.f. Figure 9), the target points at a depth of 12 m are highly affected. Small footprints of the system improve the resolution of the measurement and magnify the influence of in-water path errors. Thus, the calibration of the in-water path is crucial for object detection, especially for targets in deep water. In the absence of wave correction and propagation velocity correction, the Mapper4000U can determine the existence of a 1-m target cube and the rough shape of a 2-m target cube at a depth of 12 m.

It is worth mentioning that the reflectance of the object is also crucial for detection because it is directly related to the received power of the system [44]. A similar target cube is used to examine the object detection capability in [45], from which it is known that the reflectance of a white metal cube is 40–45% while the seafloor reflectance is around 10–20%. As the target cubes used in this experiment are ideal, natural objects will be more difficult to detect due to the low reflectance.

## 6. Conclusions

In this study, a new lightweight UAV-borne ALB system is presented and evaluated. The system weighs less than 5 kg and can be easily mounted on a multi-copter UAV platform like *DJI* Matrice 600 Pro. The system equipped with a dual-wavelength laser can flexibly measure shallow waters in small areas at a pulse repetition frequency of 4 kHz and a scanning speed of 15 lines/s.

To assess the system performance, we conducted a flight test at Dazhou Island, China, and acquired field data of two strips. From the experimental results, the main conclusions are as follows.

1. The system can simultaneously acquire land, water surface, and water bottom point clouds with a maximum detectable depth of 1.7–1.9 SD.
2. The accuracy of the system is evaluated from two aspects, water surface, and bottom. The RMSE of the water surface and bottom heights are 0.1227 m and 0.1268 m, respectively. The detection of the surface signal may be influenced by the water column backscattering, which may also be one of the reasons for the overestimation of water depths. Affected by the calculation error of the in-water path of the laser pulse, errors of water bottom points are dependent on water depths.
3. Compared to the manned ALB system, this system is lighter and more flexible and can preserve more detailed topographic features with 110 times the point density of the Mapper5000.
4. For object detection, the system can successfully detect white fabric cubes at a depth of 12 m (beyond 1 SD). The presence of the 1-m target cube and the general shape of the 2-m target cube can be observed in the point cloud. However, shape deformations of the targets also can be observed because of the depth-dependent errors, and the possibility of an object being detected is affected by its reflectance.

The experimental results have demonstrated the measurement accuracy, bathymetric performance, and object detection capability of the Mapper4000U. However, there are still some aspects that need to be further researched. For the hardware, the sampling rate of the receiver and the repetition frequency of the laser are expected to be increased, so the accuracy of signal detection and point density can be improved. In terms of the data processing software, the correction of the water column backscattering effects and the calibration of the in-water path need to be added to the data processing procedure to enhance the detectability of small objects in deep waters.

**Author Contributions:** Conceptualization, S.X., Y.H. and Q.X.; methodology, D.W., S.X., Y.H. and J.Y.; software, D.W. and J.Y.; validation, D.W., J.Y., and P.L.; formal analysis, D.W., S.X., Y.H. and J.Y.; investigation, D.W. and S.X.; writing—original draft preparation, D.W.; writing—review and editing, D.W., S.X., Y.H., J.Y., Q.X. and P.L.; visualization, D.W. and P.L.; project administration, S.X. and Y.H.; funding acquisition, S.X., Y.H., J.Y. and Q.X. All authors have read and agreed to the published version of the manuscript.

**Funding:** This research was funded by the National Natural Science Foundation of China (No. 41876105, No. 42106180, and No. 41371436), and the Foundation of Science and Technology on Near-Surface Detection Laboratory (No. TCGZ2017A008).

**Acknowledgments:** The authors acknowledge Zhenge Qiu and Weidong Zhu, Shanghai Ocean University, for the support of aerial images.

**Conflicts of Interest:** The authors declare no conflict of interest.

## References

1. Guenther, G.C.; Cunningham, A.G.; Laroque, P.E.; Reid, D.J. Meeting the Accuracy Challenge in Airborne Lidar Bathymetry. In Proceedings of the 20th EARSeL Symposium: Workshop on Lidar Remote Sensing of Land and Sea, Dresden, Germany, 16–17 June 2000.
2. Lague, D.; Feldmann, B. Topo-bathymetric airborne LiDAR for fluvial-geomorphology analysis. In *Developments in Earth Surface Processes*; Elsevier: Amsterdam, The Netherlands, 2020; Volume 23, pp. 25–54. [CrossRef]

3. Costa, B.M.; Battista, T.A.; Pittman, S.J. Comparative evaluation of airborne LiDAR and ship-based multibeam SoNAR bathymetry and intensity for mapping coral reef ecosystems. *Remote Sens. Environ.* **2009**, *113*, 1082–1100. [CrossRef]
4. Yunus, A.P.; Dou, J.; Song, X.; Avtar, R. Improved bathymetric mapping of coastal and lake environments using Sentinel-2 and Landsat-8 images. *Sensors* **2019**, *19*, 2788. [CrossRef] [PubMed]
5. Liu, Y.; Deng, R.; Li, J.; Qin, Y.; Xiong, L.; Chen, Q.; Liu, X. Multispectral bathymetry via linear unmixing of the benthic reflectance. *IEEE J. Sel. Top. Appl. Earth Observ. Remote Sens.* **2018**, *11*, 4349–4363. [CrossRef]
6. Kasvi, E.; Salmela, J.; Lotsari, E.; Kumpula, T.; Lane, S.N. Comparison of remote sensing based approaches for mapping bathymetry of shallow, clear water rivers. *Geomorphology* **2019**, *333*, 180–197. [CrossRef]
7. Agrafiotis, P.; Skarlatos, D.; Georgopoulos, A.; Karantzalos, K. DepthLearn: Learning to correct the refraction on point clouds derived from aerial imagery for accurate dense shallow water bathymetry based on SVMs-fusion with LiDAR point clouds. *Remote Sens.* **2019**, *11*, 2225. [CrossRef]
8. Fernandez-Diaz, J.C.; Glennie, C.L.; Carter, W.E.; Shrestha, R.L.; Sartori, M.P.; Singhania, A.; Legleiter, C.J.; Overstreet, B.T. Early results of simultaneous terrain and shallow water bathymetry mapping using a single-wavelength airborne LiDAR sensor. *IEEE J. Sel. Top. Appl. Earth Observ. Remote Sens.* **2014**, *7*, 623–635. [CrossRef]
9. Okhrimenko, M.; Hopkinson, C. A simplified end-user approach to lidar very shallow water bathymetric correction. *IEEE Geosci. Remote Sens. Lett.* **2019**, *17*, 3–7. [CrossRef]
10. Dreier, A.; Janßen, J.; Kuhlmann, H.; Klingbeil, L. Quality analysis of direct georeferencing in aspects of absolute accuracy and precision for a UAV-based laser scanning system. *Remote Sens.* **2021**, *13*, 3564. [CrossRef]
11. Vélez-Nicolás, M.; García-López, S.; Barbero, L.; Ruiz-Ortiz, V.; Sánchez-Bellón, Á. Applications of unmanned aerial systems (UASs) in hydrology: A review. *Remote Sens.* **2021**, *13*, 1359. [CrossRef]
12. Riegl. Riegl BDF-1 Datasheet. 2019. Available online: http://www.riegl.com/uploads/tx_pxpriegldownloads/RIEGL_BDF-1_Datasheet_2019-05-31.pdf (accessed on 15 November 2021).
13. ASTRALiTe. ASTRALiTe Edge. 2020. Available online: https://www.astralite.net/edgelidar (accessed on 15 November 2021).
14. ASTRALite. ASTRALiTe Demonstrates Scanning Topo–Bathy LiDAR System on DJI Matrice 600 Pro. 2018. Available online: https://www.businesswire.com/news/home/20181119005609/en/ASTRALiTe-Demonstrates-Scanning-Topo%E2%80%93Bathy-LiDAR-System-on-DJI-Matrice-600-Pro (accessed on 15 November 2021).
15. Mandlburger, G.; Pfennigbauer, M.; Wieser, M.; Riegl, U.; Pfeifer, N. Evaluation of a novel UAV-borne topo-bathymetric laser profiler. *Int. Arch. Photogramm. Remote Sens. Spat. Inf. Sci.* **2016**, *XLI-B1*, 933–939. [CrossRef]
16. Fuchs, E.; Mathur, A. Utilizing circular scanning in the CZMIL system. In *Algorithms Technol. Multispectral, Hyperspectral, Ultraspectral Imagery XVI, Proceedings of the SPIE Defense, Security, and Sensing, Orlando, FL, USA, 5–9 April 2010*; SPIE: Bellingham, WA, USA, 2010; Volume 7695, p. 76950W. [CrossRef]
17. Fugro. RAMMS Flyer. 2020. Available online: https://www.fugro.com/Widgets/MediaResourcesList/MediaResourceDownloadHandler.ashx?guid=eebbbdf2-f3db-6785-9f9d-ff250019aa6e&culture=en (accessed on 15 November 2021).
18. Riegl. Riegl VQ-840-G Datasheet. 2021. Available online: http://www.riegl.com/uploads/tx_pxpriegldownloads/RIEGL_VQ-840-G_Datasheet_2021-09-01.pdf (accessed on 15 November 2021).
19. Fugro. Fugro's New RAMMS Technology Advances Bathymetric Lidar Mapping Capabilities. 2018. Available online: https://lidarmag.com/2018/08/09/fugros-new-ramms-technology-advances-bathymetric-lidar-mapping-capabilities/ (accessed on 15 November 2021).
20. Mandlburger, G.; Pfennigbauer, M.; Schwarz, R.; Flöry, S.; Nussbaumer, L. Concept and performance evaluation of a novel UAV-borne topo-bathymetric LiDAR sensor. *Remote Sens.* **2020**, *12*, 986. [CrossRef]
21. Islam, M.T.; Yoshida, K.; Nishiyama, S.; Sakai, K.; Tsuda, T. Characterizing vegetated rivers using novel unmanned aerial vehicle-borne topo-bathymetric green lidar: Seasonal applications and challenges. *River Res. Appl.* **2021**, *38*, 44–58. [CrossRef]
22. Kinzel, P.J.; Legleiter, C.J.; Grams, P.E. Field evaluation of a compact, polarizing topo-bathymetric lidar across a range of river conditions. *River Res. Appl.* **2021**, *37*, 531–543. [CrossRef]
23. Pe'eri, S.; Morgan, L.V.; Philpot, W.D.; Armstrong, A.A. Land-water interface resolved from airborne LiDAR bathymetry (ALB) waveforms. *J. Coast Res.* **2011**, *62*, 75–85. [CrossRef]
24. Zhao, X.; Wang, X.; Zhao, J.; Zhou, F. An improved water-land discriminator using laser waveform amplitudes and point cloud elevations of airborne LIDAR. *J. Coast Res.* **2021**, *37*, 1158–1172. [CrossRef]
25. Mandlburger, G.; Pfeifer, N.; Soergel, U. Water Surface Reconstruction in Airborne Laser Bathymetry from Redundant Bed Observations. In Proceedings of the ISPRS Annals of Photogrammetry, Remote Sensing & Spatial Information Sciences, Wuhan, China, 8–22 September 2017. [CrossRef]
26. Zhao, J.; Zhao, X.; Zhang, H.; Zhou, F. Shallow water measurements using a single green laser corrected by building a near water surface penetration model. *Remote Sens.* **2017**, *9*, 426. [CrossRef]
27. Xing, S.; Wang, D.; Xu, Q.; Lin, Y.; Li, P.; Jiao, L.; Zhang, X.; Liu, C. A depth-adaptive waveform decomposition method for airborne LiDAR bathymetry. *Sensors* **2019**, *19*, 5065. [CrossRef]
28. Fuchs, E.; Tuell, G. Conceptual design of the CZMIL data acquisition system (DAS): Integrating a new bathymetric lidar with a commercial spectrometer and metric camera for coastal mapping applications. In *Algorithms Technol. Multispectral, Hyperspectral, Ultraspectral Imagery XVI, Proceedings of the SPIE Defense, Security, and Sensing, Orlando, FL, USA, 5–9 April 2010*; SPIE: Bellingham, WA, USA, 2010; Volume 7695, p. 76950U. [CrossRef]

29. Guth, P.L.; Van Niekerk, A.; Grohmann, C.H.; Muller, J.-P.; Hawker, L.; Florinsky, I.V.; Gesch, D.; Reuter, H.I.; Herrera-Cruz, V.; Riazanoff, S.; et al. Digital elevation models: Terminology and definitions. *Remote Sens.* **2021**, *13*, 3581. [CrossRef]
30. Yu, J.; Lu, X.; Tian, M.; Chan, T.; Chen, C. Automatic extrinsic self-calibration of mobile LiDAR systems based on planar and spherical features. *Meas. Sci. Technol.* **2021**, *32*, 065107. [CrossRef]
31. Guenther, G.C.; LaRocque, P.E.; Lillycrop, W.J. Multiple Surface Channels in Scanning Hydrographic Operational Airborne Lidar Survey (SHOALS) Airborne LiDAR. In Proceedings of the Ocean Optics XII. International Society for Optics and Photonics, Bergen, Norway, 13–15 June 1994; Volume 2258, pp. 422–430. [CrossRef]
32. Saylam, K.; Brown, R.A.; Hupp, J.R. Assessment of depth and turbidity with airborne Lidar bathymetry and multiband satellite imagery in shallow water bodies of the Alaskan North Slope. *Int. J. Appl. Earth Obs. Geoinf.* **2017**, *58*, 191–200. [CrossRef]
33. Wagner, W.; Roncat, A.; Melzer, T.; Ullrich, A. Waveform analysis techniques in airborne laser scanning. *Int. Arch. Photogramm. Remote Sens.* **2007**, *36*, 413–418.
34. Höfle, B.; Vetter, M.; Pfeifer, N.; Mandlburger, G.; Stötter, J. Water surface mapping from airborne laser scanning using signal intensity and elevation data. *Earth Surf. Processes Landf.* **2009**, *34*, 1635–1649. [CrossRef]
35. Yang, F.; Su, D.; Yue, M.; Feng, C.; Yang, A.; Wang, M. Refraction correction of airborne LiDAR bathymetry based on sea surface profile and ray tracing. *IEEE Trans. Geosci. Remote Sens.* **2017**, *55*, 6141–6149. [CrossRef]
36. Su, D.; Yang, F.; Ma, Y.; Wang, X.; Yang, A.; Qi, C. Propagated uncertainty models arising from device, environment, and target for a small laser spot airborne lidar bathymetry and its verification in the South China Sea. *IEEE Trans. Geosci. Remote Sens.* **2019**, *58*, 3213–3231. [CrossRef]
37. Birkebak, M.; Eren, F.; Pe'eri, S.; Weston, N. The effect of surface waves on airborne lidar bathymetry (ALB) measurement uncertainties. *Remote Sens.* **2018**, *10*, 453. [CrossRef]
38. Schwarz, R.; Pfeifer, N.; Pfennigbauer, M.; Mandlburger, G. Depth measurement bias in pulsed airborne laser hydrography induced by chromatic dispersion. *IEEE Geosci. Remote Sens. Lett.* **2020**, *18*, 1332–1336. [CrossRef]
39. Alvarez, L.V.; Moreno, H.A.; Segales, A.R.; Pham, T.G.; Pillar-Little, E.A.; Chilson, P.B. Merging unmanned aerial systems (UAS) imagery and echo soundings with an adaptive sampling technique for bathymetric surveys. *Remote Sens.* **2018**, *10*, 1362. [CrossRef]
40. Chowdhury, E.H.; Hassan, Q.K.; Achari, G.; Gupta, A. Use of bathymetric and LiDAR data in generating digital elevation model over the lower Athabasca River watershed in Alberta, Canada. *Water* **2017**, *9*, 19. [CrossRef]
41. Pan, Z.; Glennie, C.L.; Fernandez-Diaz, J.C.; Legleiter, C.J.; Overstreet, B. Fusion of LiDAR orthowaveforms and hyperspectral imagery for shallow river bathymetry and turbidity estimation. *IEEE Trans. Geosci. Remote Sens.* **2016**, *54*, 4165–4177. [CrossRef]
42. Genchi, S.A.; Vitale, A.J.; Perillo, G.M.E.; Seitz, C.; Delrieux, C.A. Mapping topobathymetry in a shallow tidal environment using low-cost technology. *Remote Sens.* **2020**, *12*, 1394. [CrossRef]
43. Lubczonek, J.; Kazimierski, W.; Zaniewicz, G.; Lacka, M. Methodology for combining data acquired by unmanned surface and aerial vehicles to create digital bathymetric models in shallow and ultra-shallow waters. *Remote Sens.* **2022**, *14*, 105. [CrossRef]
44. Wang, C.K.; Philpot, W.D. Using airborne bathymetric lidar to detect bottom type variation in shallow waters. *Remote Sens. Environ.* **2007**, *106*, 123–135. [CrossRef]
45. Feygels, V.; Ramnath, V.; Smith, B.; Kopilevich, Y. Meeting the international hydrographic organization requirements for bottom feature detection using the Coastal Zone Mapping and Imaging Lidar (CZMIL). In Proceedings of the OCEANS 2016 MTS/IEEE Monterey, Monterey, CA, USA, 19–23 September 2016. [CrossRef]

sensors

MDPI

Article

# Estimating Water Transport from Short-Term Vessel-Based and Long-Term Bottom-Mounted Acoustic Doppler Current Profiler Measurements in an Arctic Lagoon Connected to the Beaufort Sea

**Chunyan Li [1,*] and Kevin Mershon Boswell [2]**

1 Department of Oceanography and Coastal Sciences, College of the Coast and Environment, Louisiana State University, Baton Rouge, LA 70803, USA
2 Institute of Environment, Florida International University, North Miami, FL 33181, USA; kevin.boswell@fiu.edu
* Correspondence: cli@lsu.edu

**Abstract:** Acoustic Doppler current profilers (ADCP) are quasi-remote sensing instruments widely used in oceanography to measure velocity profiles continuously. One of the applications is the quantification of land–ocean exchange, which plays a key role in the global cycling of water, heat, and materials. This exchange mostly occurs through estuaries, lagoons, and bays. Studies on the subject thus require that observations of total volume or mass transport can be achieved. Alternatively, numerical modeling is needed for the computation of transport, which, however, also requires that the model is validated properly. Since flows across an estuary, lagoon, or bay are usually non-uniform and point measurements will not be sufficient, continuous measurements across a transect are desired but cannot be performed in the long run due to budget constraints. In this paper, we use a combination of short-term transect-based measurements from a vessel-mounted ADCP and relatively long-term point measurements from a moored ADCP at the bottom to obtain regression coefficients between the transport from the vessel-based observations and the depth-averaged velocity from the bottom-based observations. The method is applied to an Arctic lagoon by using an ADCP mounted on a buoyant platform towed by a small inflatable vessel and another ADCP mounted on a bottom deployed metal frame. The vessel-based measurements were performed continuously for nearly 5 h, which was sufficient to derive a linear regression between the datasets with an $R^2$-value of 0.89. The regression coefficients were in turn applied to the entire time for the moored instrument measurements, which are used in the interpretation of the subtidal transport variations.

**Keywords:** acoustic Doppler meters; ocean current; transport to the Arctic; moving platform measurements; regression with fixed sensors

## 1. Introduction

Doppler shift is a physical phenomenon recognized about 180 years ago [1]. When a source of waves and a receiver of those waves have a relative motion, the received frequency is dependent not only on the original frequency sent out from the source but also on the relative velocity between the source and receiver [2]. This change in frequency due to the relative motion is the Doppler shift or Doppler effect. The most common Doppler shift phenomena include those for electromagnetic waves and acoustic waves [3].

Doppler shift has been applied in technology for instrumentation in many fields [4]. For example, Doppler radar for weather has been widely used for real-time monitoring of precipitation and severe weather, including thunderstorms and tornadoes [5–9]. The U.S. is now equipped with ~159 Next-Generation Radars (NEXRAD [10,11]) for weather service.

The high frequency (HF) radar [12–16] is another example for ocean surface current measurements. There are about 150 HF radars along the coast of the U.S., including the

Great Lakes, and the data are reported in real-time to the U.S. Integrated Ocean Observing System [17]. These systems provide continuous coverage of surface flows in regions of interest for data used for research and forecast purposes.

To measure the vector field of the flow, a single wave source (the transducer) is not enough. A Doppler radar with a single wave source can only measure the speed in the radial direction but not the velocity vector because it cannot measure the velocity component in the crossbeam direction. By using two or more wave sources (transducers), a velocity vector field can be resolved as the crossbeam velocity component can be accounted for by the other transducer(s). This is achieved in the weather radar [18] and HF Radar [13] applications by overlapping the area of coverage of two adjacent radar units.

Doppler shift has also been used in acoustic waves for flow velocity measurements in the ocean. Started in the 1980s [19], commercial acoustic Doppler current profilers (ADCPs) have been applied to measure vertical profiles of 3-D velocity vectors (with east, north, and vertical components) using three to nine transducers (or beams) integrated into an underwater enclosure. The multi-beam transducers are designed to have slightly different angles allowing a mathematical solution (a matrix inversion) to compute the velocity vector. More transducers (>3) allow extra degrees of freedom for more information such as error estimates as well as for increased accuracy.

ADCPs have been widely used in measurements of ocean current velocity or suspended sediment profiles either along the vertical [20] or along the horizontal [21,22]. They can be mounted on fixed platforms, deployed on moorings on the bottom or used on moving platforms [23–26] (ships or other automated, tethered, or remotely controlled, or programmed survey vehicles). One of the applications of ADCPs is the measurements of water transport across a waterway between different regions in the ocean or an estuary [23–25]. The cross-sectionally integrated transport of water, salt, suspended sediments, nutrients, and other bio-geo-chemical materials is of great importance in studying the land-ocean exchange, the effect of climate change, and anthropogenic impact on the coastal and global environment.

Global climate change has shown a greater magnitude of trend of warming in the Arctic region [27–29]. As a result, more freshwater is transported from the permanent ice in the ocean and on land to the Beaufort Sea [30] and other regions in the Arctic Ocean. The study of this requires a combined effort using satellite remote sensing, numerical experiments, and in situ observations of water transport between the Arctic land and Arctic Ocean through rivers, estuaries, and lagoons, which are abundant in the region and yet have been the least measured due to the logistical challenges of surveying in the coastal Arctic.

The present study is motivated by this larger and longer scope of research, using ADCPs to quantify the cross-sectional water transport through an Arctic lagoon. The work alone is not able to reach the ultimate goal of the evaluation of the impact of climate change and Arctic warming. Rather, it is aimed at the development of an effective method for conducting reliable measurements using acoustic sensors. One of the challenges of the transport measurement is that the spatial coverage and temporal coverage cannot be satisfied at the same time. As shown in Figure 1, a common practice of long-term measurements of transport of water across a waterway is to use bottom-mounted ADCP(s). Because the velocity field is often non-uniform, e.g., with greater velocity in the center of the channel and smaller velocity over the shallow shoals [23–25,31,32], it is impossible to have reliable cross-sectional transport represented by 1 or 2 bottom-mounted ADCP(s). At most, they provide a proxy, but the error is difficult to obtain without additional information.

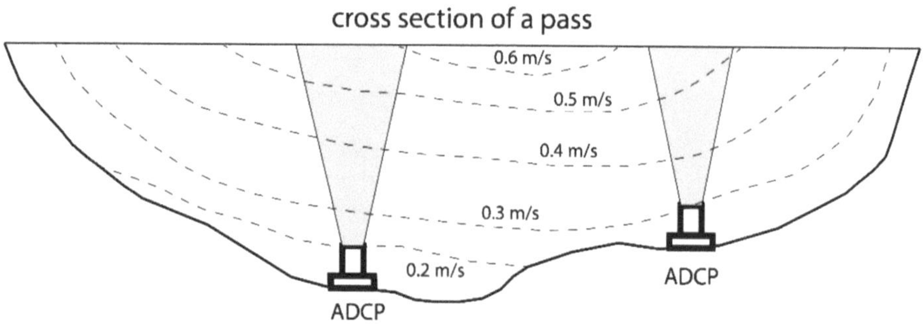


**Figure 1.** Using bottom-mounted ADCP(s) to measure flow profiles and estimate cross-sectional transport. Contour lines are hypothetical flow velocity magnitudes across the section (flow is into or out of the plane).

This leads to the design of observations during which measurements are performed simultaneously from a vessel-based ADCP transecting along a cross sectional line and a bottom-mounted ADCP deployed at a location representative of the flow inside the channel. A statistical relationship is then established between the two datasets such that the bottom-mounted ADCP data can be used to rescale to the total transport for a longer time period. This paper will introduce such a method applied to a tidal inlet of an Arctic lagoon (Elson Lagoon). The next section discusses the details of the method, the implementation of the measurements, followed by Section 3 on the major results, Section 4 with discussion, and Section 5 for the conclusions.

## 2. Materials and Methods

### 2.1. Study Site

Our experiment was conducted in Elson Lagoon in northern Alaska. It is the northernmost coastal lagoon-estuary of the United States. The northwestern portion of this system is a rectangle of ~8 × 25 km and the mean water depth is about 2–3 m (Figure 2). The lagoon is located near the confluence of the Chukchi and Beaufort Seas. It is roughly bounded within 156°36′ W, 155°54′ W, 71°12′ N, and 71°23′ N and oriented in the northwest–southeast direction. Eluitkak Pass, in the northwestern corner of the lagoon, is a relatively wide (~300 m) and deep (~16 m) channel and is where we deployed the bottom-mounted ADCP and performed the vessel-based transects. A chain of islands located east and southeast of the Eluitkak Pass is along the coast as the seaward boundary of the lagoon connecting to the coastal ocean and the Beaufort Sea. This is a region strongly influenced by the Arctic lows and highs of air pressure systems and severe storms [33].

### 2.2. Instruments

In this study, we used two ADCPs. One of them was deployed on the bottom of the Eluitkak Pass. The second one was mounted on a fiberglass surface craft. This surface craft carrying the ADCP was towed on the starboard side of an inflatable boat (Figure 3) measuring the flow velocity profiles and the cross-channel total transport at Eluitkak Pass.

The bottom-mounted instrument was a 1200 kHz Teledyne RD Instrument Workhorse ADCP, which has four beams with a Janus configuration. The vessel-mounted system was a Sontek multi-frequency M9 ADCP configured with 9-beams, four of them working at 2 MHz frequency while four other beams working at 1 MHz. The last beam was the vertical one working at 0.5 MHz to measure the water depth. To record the position of the boat, a Garmin GPSmap 60CSx was used with differential GPS.

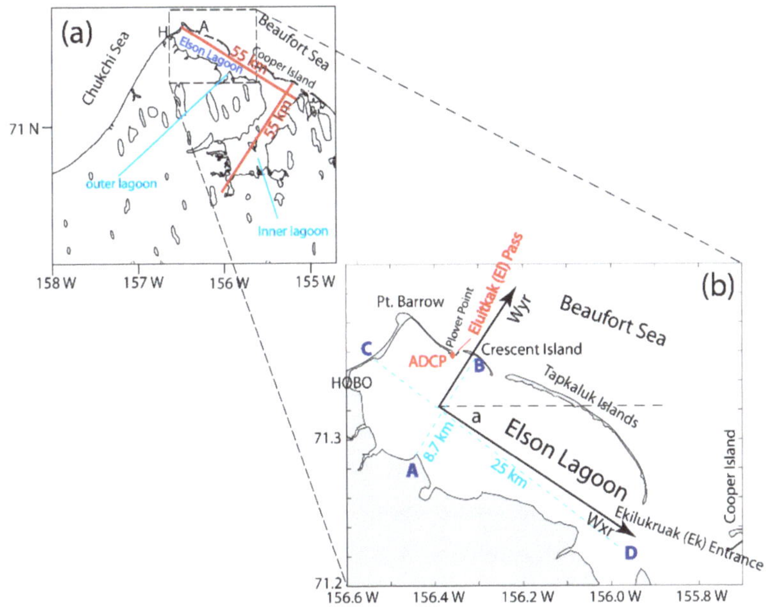


**Figure 2.** Study area at the northwestern Alaska between the Chukchi Sea and Beaufort Sea (**a**). The zoomed-in view of the Elson Lagoon (**b**) which also shows the scales and the location of bottom-mounted ADCP at the Plover Point at Eluitkak Pass.

**Figure 3.** The inflatable vessel used to tow the fiberglass surface craft (in front of the inflatable vessel), which carried the M9 ADCP.

### 2.3. Measurements

The mooring was comprised of an aluminum cross with 5 pounds of extra weight on each of the four "legs" for stability, on which the ADCP was mounted in an upward direction. The ADCP's compass was calibrated before the deployment according to the manufacturer's procedures. The ADCP's internal clock was set to record in UTC. It was deployed at (71.3593° N, 156.3561° W, on the western side of the inlet) at a depth of about 13.35 m, on 29 July 2014. With a blanking distance of ~1 m, the bottom most data point was at 1.53 m and the vertical interval of measurement was 1.00 m. The ADCP was setup to sample once every 80 s and 45 times per hour (Table 1). The averaged hourly data were saved in the internal memory. The starting time of valid data was 1630 UTC 29 July 2014 and the last valid ensemble of measurements was at 0230 UTC, 3 August 2014 for the first deployment. The second deployment of this ADCP was made about one day later with the valid data between 0320 UTC 4 August 2014, and 1800 UTC, 13 August 2014. The instrument was deployed at a slightly different location (71.3597° N, 156.3538° W), northeast of the first location at a depth of 11.00 m. The second deployment also had a different setup for the sampling schemes. Specifically, the vertical intervals were changed to 0.25 m (instead of 1.00 m) and the ensemble sampling time interval was set to 5 min, within which 50 samples would be taken at 6 s intervals for the 5 min ensemble average (Table 1).

**Table 1.** ADCP parameters.

| ADCP | Bottom Mounted 1 | Bottom Mounted 2 | Vessel Based M9 |
|---|---|---|---|
| Raw sampling interval (s) | 80 | 6 | variable (~1/7–1/40) |
| Ensemble interval | 1 h | 5 min | 1 s |
| Vertical bin size (m) | 1 | 0.25 | variable |

The M9 ADCP's compass was calibrated according to the manufacturer's suggested procedures. The internal clock of the M9 was set to record in UTC. About seven hours after the bottom-mounted ADCP was deployed, just prior to 1630 UTC on 29 July 2014, the inflatable vessel towing the M9 ADCP started to run across the ~300 m wide Eluitkak Pass. For the entire measurement period, the M9 ADCP was operating under the "bottom tracking mode", which used the strong bottom return signal originally sent from the ADCP transducers to compute the instantaneous velocity of the vessel relative to the sea bottom. This bottom tracking velocity was then used to compute the Earth-coordinate velocity components for the water particles. This approach gives a higher accuracy for the velocity field of the water, compared to using the raw GPS data for the same purpose: if the real-time raw GPS data were used to compute the velocity of the vessel, it would have introduced a much greater error in the water velocity computation.

Data collection by the M9 ADCP commenced at about 2322 UTC. This was performed concurrently with the Garmin GPS recording data at 1 s intervals. The inflatable vessel was running across the Eluitkak Pass repeatedly for 41 times until 0414 UTC 30 July 2014. During this time period, the average speed of the boat was estimated at about 0.7 m/s. The transect across Eluitkak Pass passed the bottom-mounted ADCP so the flow velocities from both boat-based and bottom deployed ADCPs can be directly compared.

Note that it was a challenge and risky operation to run an inflatable boat in that remote area. The fog, winds, waves, and cold air and water make it difficult to maintain the transect lines in the small inflatable vessel; however, in consideration of these conditions, the survey and data collection were quite successful.

### 2.4. Data Processing

The M9 ADCP data included accurate time stamps but not the geo-location. With the time series of vessel positions from the Garmin GPS, we merged the GPS data with the ADCP data by simple interpolations. Since both M9 and Garmin GPS recorded data at 1-s intervals, they were comparable in time increments, and the interpolation maintained the quality of the data.

The M9 ADCP data provide outputs for both velocity profiles and integrated total transport across the channel. To extract the total transport, the start and end points of each transect must be determined. This was completed for all the 41 repetitions across the channel (Figure 4). Among the 41 transects, we selected 37, excluding 4 that were too far away from the intended transect. The water depth measured from the vessel-based ADCP is shown in Figure 5. The main channel is on the western end, and the eastern end has a relatively wider shoal of 2–5 m.

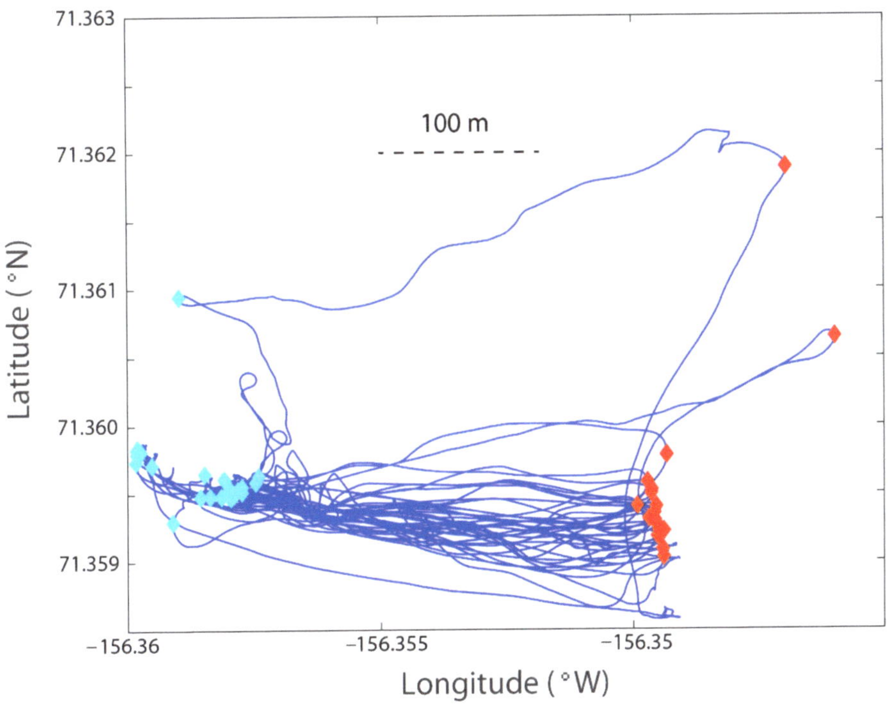


**Figure 4.** Vessel track and points selected for the west and east ends of a transect for total cross-sectionally integrated transport. The light blue and red diamonds represent the west and east end of the transects, respectively. Four of the forty-one lines were excluded in the analysis as they were too far away from the intended transect. A horizontal scale of 100 m is indicated.

To extract the velocity data from the vessel-based ADCP to compare with those from the bottom-mounted ADCP, we defined a rectangle around the deployed ADCP (Figure 5, green box) with a dimension of roughly 74 m in the north–south direction and 47 m in the east–west direction. This 74 × 47 m data footprint (the green box) is a proper selection as the water depth within this box is consistent (varying within 2 m) and the vessel's speed (averaged ~0.7 m/s) is slow enough to permit sufficient sampling inside the box for a reliable statistical averaged velocity each time the vessel passed through the box.

The middle (average) time of each of the transects was used to interpolate the bottom ADCP data onto the same time of the vessel-based transport measurements (37 transport values). This was performed to create a time series for the velocity vector at each of the vertical locations, as well as the depth-averaged velocity. This permits the development of a regression between the velocity from the bottom-mounted ADCP and the boat-based transport measurements:

$$P = \alpha_1 v + \alpha_2 \tag{1}$$

in which $P$ is the cross-channel volume transport measured from the vessel, $v$ is the north component of the velocity measured by the bottom-mounted ADCP, and $\alpha_1$ and $\alpha_2$ are the regression coefficients. Note that different sets of $\alpha_1$ and $\alpha_2$ are expected for velocities from different vertical positions and for the depth-averaged velocity. The east component of the velocity from the bottom ADCP was not used as it was roughly in the cross-channel direction and not correlated with the transport.

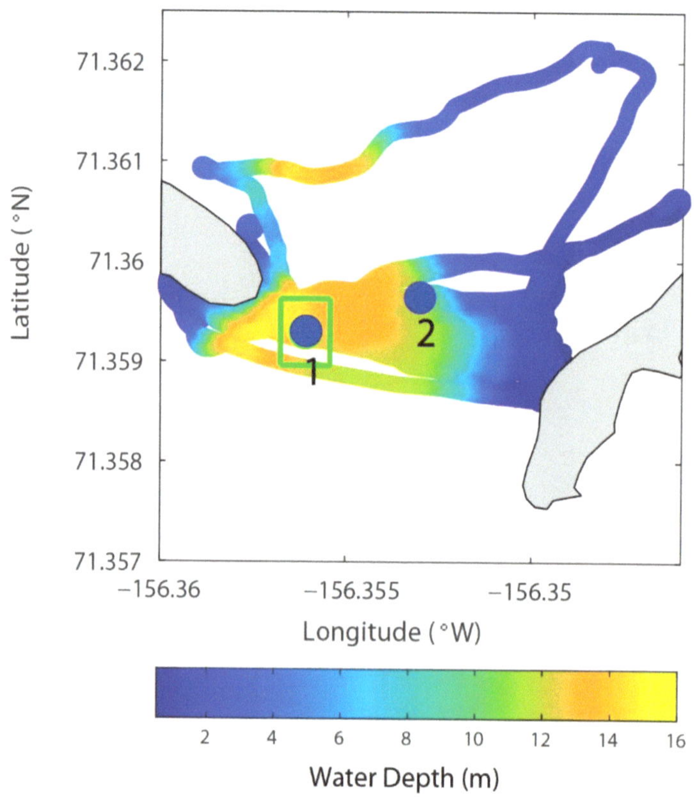


**Figure 5.** ADCP locations and water depth measured by the vessel-based M9 ADCP. The locations of the bottom-mounted ADCP are shown by the blue circles and the numbers (1 and 2) indicate the first and second deployments, respectively. The green rectangular box shows the region selected for capturing the flow velocity data from the vessel-based ADCP to compare with the flow data from the first mooring.

If the correlation between the two components is high, we can use the regression coefficients to compute the transport from the velocity time series measured by the bottom-mounted ADCP. This will be useful because the bottom-mounted ADCP was deployed for more than one week while the vessel-based survey would not be safe to run at night, and it is impossible to continue for several days, considering the challenging Arctic conditions there.

Since the second deployment was at a slightly shallower depth (~11.00 m) than the first (13.35 m), the velocity magnitude measured is affected by the depth difference. As demonstrated in [34] for the mean flow or low frequency (quasi-steady-state) flows and [32] for tidal flow, the velocity magnitude is dependent on water depth. When compared along *the same cross section*, assuming all other parameters are the same, the shallower water will have a smaller velocity magnitude for both low frequency (or mean) flow and tidal flow. Thus, we need to find the velocity factors to transform the velocity at the second site of deployment to that at the first site so the regression coefficients can be applied to compute the total cross-sectional transport for the entire time periods of both deployments. More specifically, for the low-frequency flow component [34],

$$u_1 = u_2\sqrt{\frac{h_1}{h_2}} \tag{2}$$

in which $u_1$ and $u_2$ are the velocities at site 1 and 2, with depth $h_1$ and $h_2$, respectively. In our case, $h_1 = 13.35$ m, $h_2 = 11.00$ m, so the factor for low-frequency velocity transformation from site 2 to site 1 is

$$f_1 = \sqrt{\frac{h_1}{h_2}} = 1.1017 \tag{3}$$

For the tidal flow, from [32], the factor for velocity transformation from site 2 to site 1 is

$$f_2 = \frac{h_1}{h_2}\sqrt{\frac{\sigma^2 h_1^2 + \beta^2}{\sigma^2 h_2^2 + \beta^2}} \tag{4}$$

in which $\sigma$ is the angular frequency for tide, which in this region is semi-diurnal, or

$$\sigma = \frac{2\pi}{12 \times 3600} \quad (\text{rad/s}) \tag{5}$$

and $\beta$ is a friction coefficient defined by [35,36]

$$\beta = \frac{8C_D U_0}{3\pi} \tag{6}$$

where $C_D$ and $U$ are the bottom drag coefficient and tidal velocity amplitude, respectively. In this study, we choose the typical value for the drag coefficient $C_D = 0.0025$ [36], and $U_0 = 0.5$ m/s based on our data. This yields,

$$f_2 = 1.053 \tag{7}$$

Therefore, to transform the velocity measured from site 2 to site 1, the low-frequency component should be increased by about 10% ($f_1 = 1.1017$) and the tidal component by about 5% ($f_2 = 1.053$).

To transform the velocity measured at site 2 during the second deployment to that at site 1 so we can implement the regression coefficient for the transport computation for the second deployment period, we low-pass filtered the data from the second deployment to separate the tidal and non-tidal (low frequency) velocity components. For that purpose, a 40-h Butterworth [37] low-pass filter was used for the velocity data from the second deployment to separate the time series of depth-averaged velocity $v_2$ into

$$v_2 = v_L + v_T \tag{8}$$

here, $v_L$ and $v_T$ are the low-frequency and tidal velocity components, respectively. The reason we used a 40-h cut-off for the low-pass filter is because we need to filter out diurnal tidal constituents. Although tide in the region is basically semi-diurnal, there is diurnal inequality and thus diurnal constituents are not exactly zero. Using a 40-h cut-off can eliminate any diurnal tide and eliminate any sidelobe leakage effect.

By applying the factor $f_1$ to $v_L$ and $f_2$ to $v_T$, an approximated velocity at site 1 for the second period can be obtained before applying the regression coefficients to obtain the total transport:

$$v_1 = f_1 v_L + f_2 v_T \tag{9}$$

The total transport is then

$$P = (v_1 \; I)\begin{pmatrix} \alpha_1 \\ \alpha_2 \end{pmatrix} \tag{10}$$

here, $I$ is an array of 1′s of $N \times 1$ dimension, in which $N$ is the length of the time series $v_1$ or $v_2$ (same length), and $\alpha_1$ and $\alpha_2$ are the regression coefficients in (1).

## 3. Results

### *3.1. Velocity Comparison*

The velocity measured from the bottom-mounted ADCP during the transect measurement varied between 0.3 and 0.55 m/s. The vessel-based ADCP measured greater velocity variance. This was expected because the transects covered a much larger area, including the shallow waters on the eastern end and the deepest channel (~16 m), which was slightly northwest of the bottom deployed ADCP at site 1 (within 30–50 m range, Figure 5). Nevertheless, the velocity from these two ADCPs showed consistency (Figure 6).

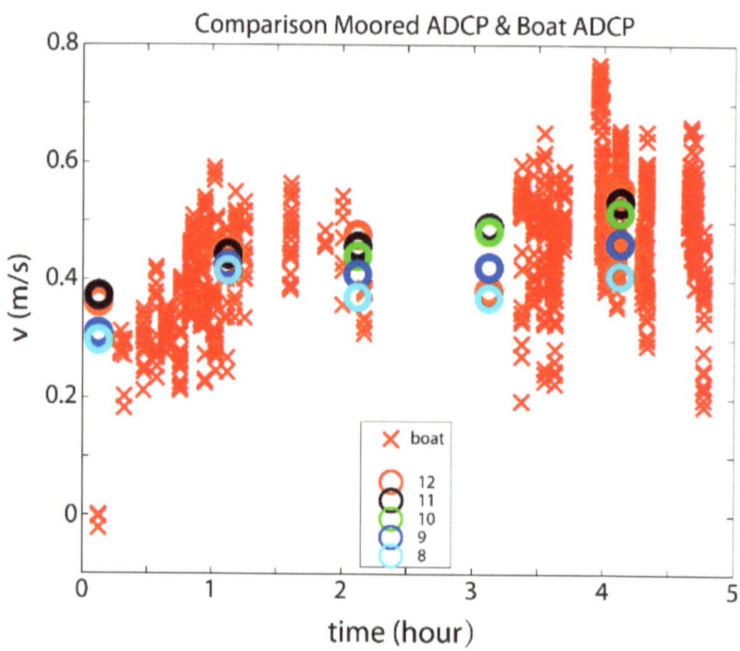


**Figure 6.** Comparison between velocities. Red crosses are the north velocity component from the vessel-based ADCP at different depths. The colored circles are the velocity data from the bottom-mounted ADCP at different heights above the bottom (e.g., the red circle is 12 m above the bottom, black circle is 11 m above the bottom, etc.).

### *3.2. Regression of Transport*

The horizontal velocity showed a dependency on vertical position: the nearer the bottom, the lower the velocity. As the depth-averaged velocity varied from 0.3 to 0.45 m/s, the transport tripled from 400 to 1200 m$^3$/s. Compared with the regression with velocities at different vertical positions, the depth-averaged velocity had the best correlation with the total transport from the vessel-based measurements with an $R^2$ value of 0.89 (Figure 7). The coefficients $\alpha_1$ and $\alpha_2$ were 4801.9 and −971.8, respectively. The 95% confidence intervals for these two parameters were (4214.6, 5389.3) and (−1200.7, −742.9), respectively. From this result, we can see that the correlation between the two instruments is significant. The negative value for $\alpha_2$ indicates that there is an inward transport when the depth-averaged velocity from the bottom-mounted ADCP site is 0. This may imply that there is an inward flow in the shallow waters away from the ADCP site when the flow in the channel is small.

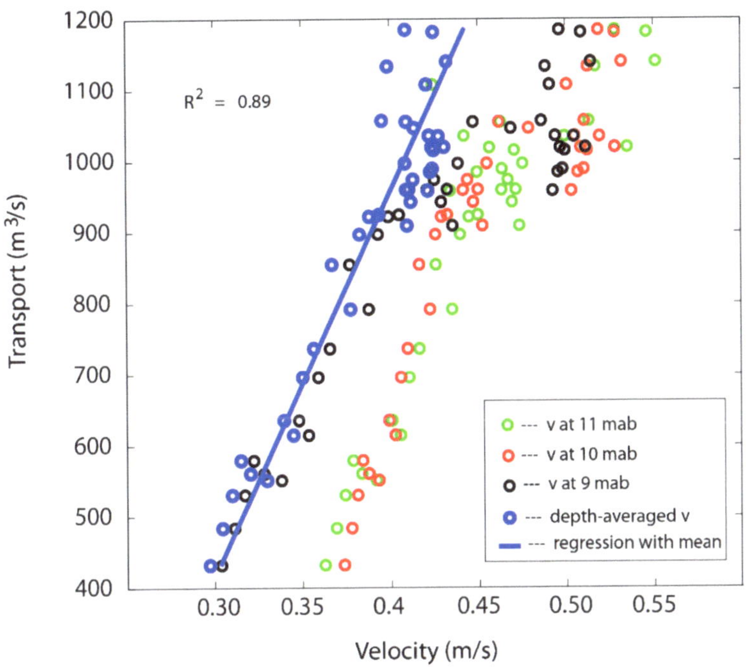


**Figure 7.** Linear regression between the velocity measured by the bottom-mounted Teledyne RDI ADCP at site 1 with the total cross channel transport measured by the vessel-based M9 ADCP during the first deployment. The circles are for velocity from the bottom-mounted ADCP at various heights (meter above the bottom, or mab). The blue circles are the depth-averaged velocity from the bottom-mounted ADCP.

### *3.3. Transport Time Series*

Using the regression coefficients $\alpha_1$ and $\alpha_2$, the transport for the entire time of the first deployment (about 4.5 days) can be calculated based on the depth-averaged velocity from the bottom-mounted ADCP at site 1. Applying Equations (8)–(10), the transport for the entire time of the second deployment (about 4 days) can also be calculated based on the depth-averaged velocity from the bottom-mounted ADCP at site 2. The maximum outward transport (positive sign means outward transport) was about 3800 $m^3/s$ during the first deployment (Figure 8). There was a strong outward flow between days 3 and 4 during the first deployment when the total transport was greater than 6000 $m^3/s$ into the lagoon (negative sign). This turns out to be a significant event because of a northwesterly wind associated with an Arctic low air pressure system (see below in the discussion section) that induced inward transport through the multiple inlet hydrodynamics [22,38] as well as Ekman transport. The second deployment period did not have such strong inward transport, but the outward transport was comparable. Furthermore, semi-diurnal tidal signals are obvious for both periods of deployment. The transport from this project showed that the non-tidal variation amounts to 10,000 $m^3/s$ (−6000 to 4000), while the tidal variations only had 30–50% of that, indicating that in this region, the wind-driven flows are more important than those of the tidal motions.

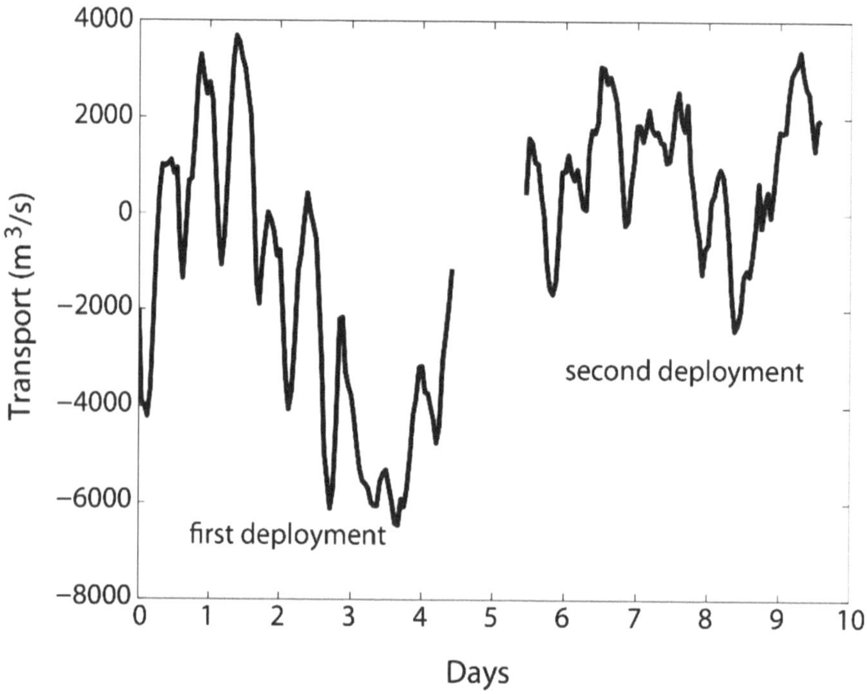


**Figure 8.** Transport computed for the first and second deployments. The horizontal axis is time in days from the beginning of the first deployment.

## 4. Discussion

### *4.1. Weather Conditions*

Our study happened to coincide with a strong atmospheric low-pressure system passing north of the region in the Beaufort Sea (Figure 9) one day before the retrieval of the ADCP during the first deployment. A cyclonic wind associated with the low atmospheric system brought a strong northwesterly wind to the Elson Lagoon (Figure 9). This produced a significant inward transport exceeding 6000 $m^3/s$ (Figure 8).

The weather data recorded by an automated surface observation system (ASOS) at the airport in Utqiaġvik showed this event. The local weather data showed that the event started with a warming of the ground level air from nearly 0 C to about 11 C in two days. The air temperature then dropped to below 0 C in another two days (upper panel of Figure 10). On 2 August (day 214 of 2014), the local sea level air pressure reached a minimum from about 1028 to 1009 mb (middle panel of Figure 10). Concurrently, wind speed increased and reached maximum almost at the same time when the air pressure reached its minimum. The wind direction was roughly northwest between Day 214 and Day 215 of 2014 (1 January was defined as Day 1). By analyzing the weather data for the whole year, the event of 2 August appeared to be the most intensive local event for the summer of 2014.

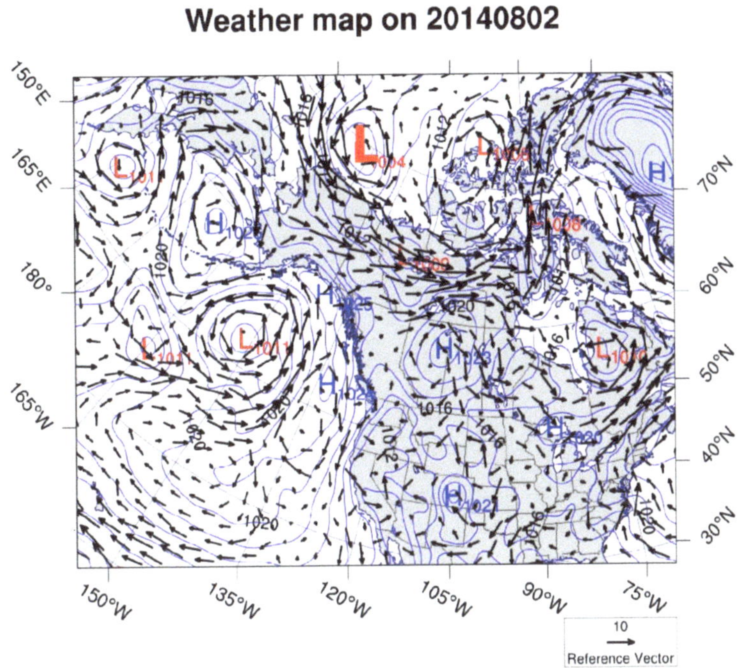


**Figure 9.** Weather map from NOAA's reanalysis data for 2 August 2014. The low air pressure system indicated by a large red "L" in the upper portion of the map indicates the wind (to the southwest of the "L") was from the northwest, suggesting Ekman transport into the Elson Lagoon.

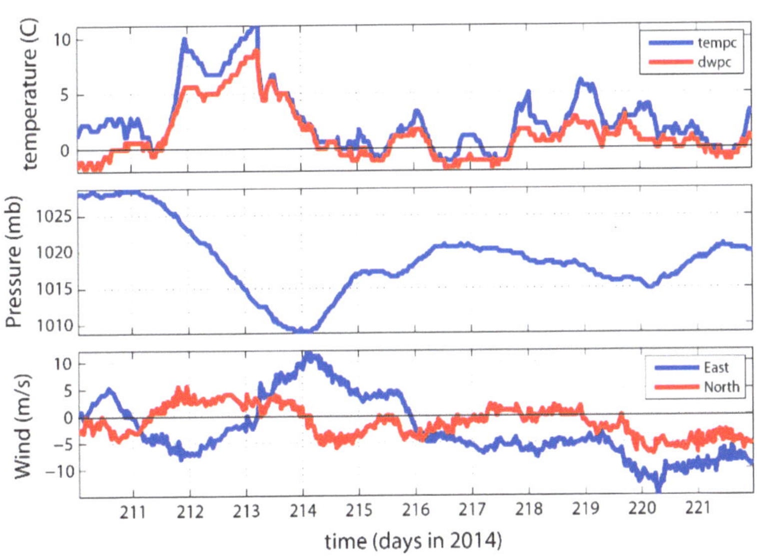


**Figure 10.** Weather time series data from the Barrow airport automated surface observation system (ASOS). The upper, middle, and lower panels are for the air temperature and dew point temperature, air pressure, and wind velocity components (east and north components); 2 August 2014 is day 214 starting from 1 January.

### 4.2. Comment about the Method

Ideally, if an ADCP can be used on a moving platform to acoustically measure the velocity profiles across a channel continuously, a reliable time series of transport can be obtained. This, however, is hardly realistic for most places because of the cost involved, not to mention the possibility in an environment with no infrastructure, and hazards associated with potential floating ice and freezing in the winter season. Alternatively, if an array of ADCPs can be deployed along the bottom, reliable measurements of transport over time is feasible. However, this is also problematic because of the cost involved to have multiple ADCPs and the risk of damage or loss in dynamic shallow water environments is very high in this region within the Arctic.

The advantage of the method presented in this paper is that it avoids conducting continuous measurements of the transport or using costly multiple ADCPs for bottom deployment. Instead, it establishes a correlation between the depth-averaged velocity from a bottom-mounted ADCP with a short-term vessel-based measurement of transport. This is particularly useful for applications in the Arctic because of the adverse environmental conditions there (low temperature, lack of facilities and even a standard boat launch).

Although the method has been shown to be valid and useful by our study, we would also like to add a few cautionary notes. The first is that the vessel-based surveys should be long enough to capture sufficient variability over time. In our study, the length was nearly 5 h and indicated a strong positive relationship. We expect that if the duration of the survey was longer (e.g., 24 h) it would be ideal as it would cover two tidal cycles. This method has been used in our previous studies in the Louisiana tidal channels [39] in which we had the luxury of having an entire diurnal tidal cycle (close to 25 h), which resulted in an even stronger relationship with $R^2$ values ranging between 0.96 and 0.99. In the Arctic region, this is a novel experiment, and the results are encouraging. The method, however, has not been discussed in detail as in this paper. The work presented in [39] only presented the results with the context of examining weather-induced exchange flows through a tidal channel in southern Louisiana. This paper provides a detailed explanation for the first time with an actual challenging application in the Arctic.

### 4.3. On the Measurement Errors

*The use of bottom-tracking*. As described earlier, we used the bottom-tracking mode for the M9 ADCP. The original relative velocity between the moving vessel and the water is measured by the Doppler shift. This also considers the pitch and roll and orientation (with the onboard IMU and compass) and gives the three components of velocity profiles along the vertical. The velocity at each of the vertical positions has three components in the three-dimensional Cartesian $xyz$ coordinate. However, that has not considered the velocity of the moving vessel. The velocity of the moving vessel is usually not calculated from the GPS because of relatively large errors from the raw GPS data—the ADCP needs to sample many times in a second (it varies with the M9, which, in our case, sampled 7-40 times per second, Table 1), and it is hard for a GPS to keep pace with the fast-updating requirements. Instead, the speed of the vessel is measured by "bottom-tracking". This is a function of the ADCP using the same Doppler shift principle to measure the velocity of the sea bottom. This is generally more accurate than using the GPS for most applications unless a very high-resolution RTK GPS is used. After the bottom tracking velocity is obtained, it is subtracted from the $xyz$ velocity to yield the Earth coordinate (ENU or East, North, Up) velocity components. The GPS unit used in this study is only for time stamping of the time series data and to link the bottom ADCP data with the vessel-based ADCP data. The position error is the general differential GPS error (3–6 m in position). However, since our footprint for the bottom ADCP is a rectangle of $74 \times 47$ m (Figure 5), the error in position is negligible.

*The Errors for ADCP Data*. Both ADCPs provide error velocity estimates for each and every ensemble sample. For example, the error velocity from the bottom-mounted ADCP ranged from $-0.03$ to $0.03$ m/s (Figure 11). The depth-averaged error velocity at each of the hours for the five hourly data are −0.0322, 0.0052, 0.0016, 0.0132, and 0.0152 m/s, respectively. The overall averaged error velocity is less than 0.01 m/s. Likewise, the error velocities of the M9 ADCP were also automatically computed by the instrument. The mean error velocity was computed to be less than 0.01 m/s with a standard deviation of ~0.07 m/s for the error. A greater range of error velocity compared to the bottom-mounted ADCP data is expected because of extra error introduced by the moving platform. This is consistent with previous studies. For example, the standard error for the velocity data from six full tidal-cycle surveys using a small research vessel in a tidal channel [40] was estimated to be between 0.09 and 0.17 m/s.

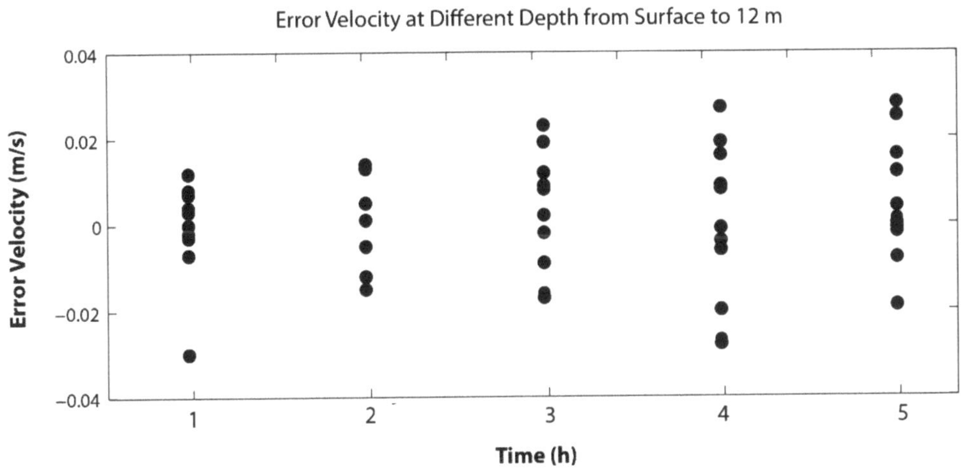


**Figure 11.** Velocity error from the bottom-mounted ADCP. The different dots at a given time are for the error velocity values at different depths from 1 to 12 m above the bottom (mab).

## 5. Concluding Remarks

The successful implementation of the proposed method in a coastal environment (e.g., tidal rivers, tidal inlets, estuaries, and straits) relies on several key factors:

(1) The time from all relevant equipment (bottom-mounted ADCP, vessel-based ADCP, and GPS) must be unified with the GPS time. UTC should be used to avoid confusion with the local time and/or daylight-saving time.

(2) The vessel-based ADCP should use the "bottom-tracking mode" unless the sea bottom is not solid (such as full of fluid mud). This in most cases will enhance the quality of the velocity data. In regions where water depth is too large and the sampling frequency of the ADCP is too high such that the ADCP could not sense the bottom, the "navigational mode" needs to be used, which in most cases might significantly increase the error of velocity measurements unless a high-resolution RTK GPS system is used. This is because of the random errors from the raw GPS, especially when high sampling rate is required for obtaining ADCP ensemble velocity values (e.g., 7-40 measurements are made to obtain a 1-s ensemble value for the M9 in our study, Table 1). In the case of using navigational mode, a temporal average of the ensemble velocity data can help in reducing the velocity error. Fortunately, this is unlikely in most coastal waters such as estuaries and lagoons because of their inherently shallow water. For example, the M9 ADCP can successfully use bottom-tracking mode in waters of 25 m. For a 600 KHz RDI ADCP, this depth can be increased to 60 m or more.

(3) The repeated measurements across the transect are very important to establish the statistical regression coefficients between the transport (from bottom-mounted ADCP)

and velocity (from the vessel-based ADCP). In general, the more repetitions, the better.

(4) The temporal length of the measurements should be "long enough" to include certain variability of the flow velocity and total cross-sectional transport. In a tidal environment, this depends on the type of tides. The time should be long enough over which the flow velocity experiences sufficient variations for obtaining a reliable statistical regression. For a semi-diurnal tidal environment, the whole tidal cycle is about 12 h, and over 3–4 h, the flow can experience 1/4 to 1/3 of the one period for tidal currents, although measurements over a complete tidal cycle is preferred if possible [39].

(5) The cross-channel transect should pass the deployed ADCP: the closer the better. In choppy conditions, this might be difficult, but with numerous repetitions, enough valid samplings can be guaranteed.

In conclusion, using a combination of a longer-term bottom-mounted ADCP (the first ADCP), measuring the local velocity profiles in a deep channel of a tidal inlet and a shorter-term boat-based ADCP (the second ADCP) measuring the cross-channel transport continuously can allow us to establish a regression between the depth-averaged velocity from the first ADCP and the transport from the second ADCP. The regression coefficients can then be applied to the longer time series from the first ADCP and obtain the transport time series from the entire deployment. This appears to be an efficient and economical way to determine the total transport. This is useful particularly considering that during severe weather, a boat-based survey is usually not possible because of safety issues, unless a reliable automated unmanned platform is used, which can also be costly and has a high risk in a remote area such as the Arctic lagoons. This method can be used in many applications in the quantification of flux of water under tidal and weather forcing. This can be particularly useful in a system with multiple inlets so that coordinated observations can be made to quantify the fluxes through different inlets, which can help the understanding of the circulation dynamics and reliable quantification of the water exchange [22,38] of the system.

**Author Contributions:** Conceptualization, C.L.; methodology, C.L. and K.M.B.; formal analysis, C.L.; investigation, C.L.; resources, C.L. and K.M.B.; data curation, C.L.; writing—original draft preparation, C.L.; writing—review and editing, C.L. and K.M.B.; visualization, C.L.; supervision, C.L.; project administration, C.L. and K.M.B.; funding acquisition, C.L. and K.M.B. All authors have read and agreed to the published version of the manuscript.

**Funding:** This research was funded by the North Pacific Research Board (NPRB, project #1229).

**Institutional Review Board Statement:** Not applicable.

**Informed Consent Statement:** Not applicable.

**Data Availability Statement:** Data for this paper are obtained by the authors in the Elson lagoon and can be found at https://oceandynamics.lsu.edu/Data.htm (accessed on 21 December 2021). Any questions should be directed to the authors.

**Acknowledgments:** Logistic support was provided by both the North Slope Borough and UMIAQ. Thanks are specifically given to Michael Donovan, who discovered and helped to retrieve the ADCP dragged by ice nearly 2 miles away on the inner shelf near the Beaufort Sea, using his own boat Jenilee. This is contribution #1388 from the Coastlines and Oceans Division of the Institute of Environment at Florida International University.

**Conflicts of Interest:** The authors declare no conflict of interest.

## References

1. Voti, R.L.; Bertolotti, M. Thermal waves emitted by moving sources and the Doppler effect. *Int. J. Heat Mass Transf.* **2021**, *176*, 121098. [CrossRef]
2. Feynman, R.P.; Leighton, R.B.; Sands, M. *The Feynman Lectures on Physics*; Volume 1: Mainly Mechanics, Radiation, and Heat; California Institute of Technology: Pasadena, CA, USA, 1963; p. 967.
3. Walker, J. *Fundamentals of Physics*, 10th ed.; Wiley: Hoboken, NJ, USA, 2013; p. 1386.

4. Alabaster, C. *Pulse Doppler Radar—Principles, Technology, Applications*; SCiTECH: Edison, NJ, USA, 2012; p. 411.
5. Bech, J.; Chau, J.L. *Doppler RADAR Observations—Weather RADAR, Wind Profiler, Ionospheric RADAR, and Other Advanced Applications*; InTech: Rijeka, Croatia, 2012; p. 470.
6. Refan, M.; Hangan, H.; Wurman, J.; Kosiba, K. Doppler radar-derived wind field of five tornado events with application to engineering simulations. *Eng. Struct.* **2017**, *148*, 509–521. [CrossRef]
7. Thiruvengadam, P.; Indu, J.; Ghosh, S. Assimilation of Doppler Weather Radar data with a regional WRF-3DVAR system: Impact of control variables on forecasts of a heavy rainfall case. *Adv. Water Resour.* **2019**, *126*, 24–39. [CrossRef]
8. Dutta, D.; Sharma, S.; Sen, G.; Kannan, B.; Venketswarlu, S.; Gairola, R.; Das, J.; Viswanathan, G.; Sharma, S.K. An Artificial Neural Network based approach for estimation of rain intensity from spectral moments of a Doppler Weather Radar. *Adv. Space Res.* **2011**, *47*, 1949–1957. [CrossRef]
9. Chen, Z.-Q.; Kavvas, M. An automated method for representing, tracking and forecasting rain fields of severe storms by Doppler weather radars. *J. Hydrol.* **1992**, *132*, 179–200. [CrossRef]
10. Xie, H.; Zhou, X.; Vivoni, E.R.; Hendrickx, J.M.; Small, E.E. GIS-based NEXRAD Stage III precipitation database: Automated approaches for data processing and visualization. *Comput. Geosci.* **2005**, *31*, 65–76. [CrossRef]
11. Krajewski, W.F.; Ntelekos, A.A.; Goska, R. A GIS-based methodology for the assessment of weather radar beam blockage in mountainous regions: Two examples from the US NEXRAD network. *Comput. Geosci.* **2006**, *32*, 283–302. [CrossRef]
12. Solabarrieta, L.; Rubio, A.; Castanedo, S.; Medina, R.; Charria, G.; Hernandez, C. Surface water circulation patterns in the southeastern Bay of Biscay: New evidences from HF radar data. *Cont. Shelf Res.* **2014**, *74*, 60–76. [CrossRef]
13. Liu, Y.; Weisberg, R.H.; Vignudelli, S.; Roblou, L.; Merz, C. Comparison of the X-TRACK altimetry estimated currents with moored ADCP and HF radar observations on the West Florida Shelf. *Adv. Space Res.* **2012**, *50*, 1085–1098. [CrossRef]
14. Chioua, J.; Dastis, C.; González, C.; Reyes, E.; Mañanes, R.; Ruiz, M.; Álvarez, E.; Yanguas, F.; Romero, J.; Bruno, M. Water exchange between Algeciras Bay and the Strait of Gibraltar: A study based on HF coastal radar. *Estuar. Coast. Shelf Sci.* **2017**, *196*, 109–122. [CrossRef]
15. Mujiasih, S.; Hartanto, D.; Beckers, J.-M.; Barth, A. Reducing the error in estimates of the Sunda Strait currents by blending HF radar currents with model results. *Cont. Shelf Res.* **2021**, *228*, 104512. [CrossRef]
16. Jeanette Romero-Cozar, J.C.; Bolado-Penagos, M.; Reyes-Perez, J.; Gomiz-Pascual, J.J.; Vazquez, A.; Sirviente, S.; Bruno, M. Tidally-induced submesoscale features in the atlantic jet and Western Alboran Gyre: A study based on HF radar and satellite images. *Estuar. Coast. Shelf Sci.* **2021**, *250*, 107122. [CrossRef]
17. Schoch, G.C.; McCammon, M. Demonstrating the Alaska Ocean Observing System in Prince William Sound. *Cont. Shelf Res.* **2013**, *63*, S149–S158. [CrossRef]
18. Insanic, E.; Siqueira, P.R. Real-Time Vector Velocity Estimation in Doppler Radar Networks. *IEEE Trans. Geosci. Remote Sens.* **2011**, *50*, 568–584. [CrossRef]
19. Teledyne, R.D. *Instruments. Acoustic Doppler Current Profiler Principles of Operation a Practical Primer*; Teledyne R.D. Instruments: Poway, CA, USA, 2011; p. 56.
20. Book, J.W.; Perkins, H.T.; Cavaleri, L.; Doyle, J.D.; Pullen, J.D. ADCP observations of the western Adriatic slope current during winter of 2001. *Prog. Oceanogr.* **2005**, *66*, 270–286. [CrossRef]
21. Moore, S.A.; le Coz, J.; Hurther, D.; Paquier, A. On the application of horizontal ADCPs to suspended sediment transport surveys in rivers. *Cont. Shelf Res.* **2012**, *46*, 50–63. [CrossRef]
22. Li, C.; Huang, W.; Milan, B. Atmospheric Cold Front–Induced Exchange Flows through a Microtidal Multi-Inlet Bay: Analysis Using Multiple Horizontal ADCPs and FVCOM Simulations. *J. Atmos. Ocean. Technol.* **2019**, *36*, 443–472. [CrossRef]
23. AlYousifa, A.; Valle-Levinson, A.; Adams, P.N.; Laurel-Castillo, J.A. Tidal and subtidal hydrodynamics over ridge-swale bathymetry. *Cont. Shelf Res.* **2021**, *219*, 104392. [CrossRef]
24. Li, C.; White, J.R.; Chen, C.; Lin, H.; Weeks, E.; Galvan, K.; Bargu, S. Summertime tidal flushing of Barataria Bay: Transports of water and suspended sediments. *J. Geophys. Res.* **2011**, 116. [CrossRef]
25. Li, C. Axial convergence fronts in a barotropic tidal inlet—Sand shoal inlet, VA. *Cont. Shelf Res.* **2002**, *22*, 2633–2653. [CrossRef]
26. Savidge, D.K.; Amft, J.A. Circulation on the West Antarctic Peninsula derived from 6 years of shipboard ADCP transects. *Deep Sea Res.* **2009**, *56*, 1633–1655. [CrossRef]
27. Alizadeh, O.; Lin, Z. Rapid Arctic warming and its link to the waviness and strength of the westerly jet stream over West Asia. *Glob. Planet. Chang.* **2021**, *199*, 103447. [CrossRef]
28. You, Q.; Cai, Z.; Pepin, N.; Chen, D.; Ahrens, B.; Jiang, Z.; Wu, F.; Kang, S.; Zhang, R.; Wu, T.; et al. Warming amplification over the Arctic Pole and Third Pole: Trends, mechanisms and consequences. *Earth-Sci. Rev.* **2021**, *217*, 103625. [CrossRef]
29. Kumar, A.; Yadav, J.; Mohan, R. Global warming leading to alarming recession of the Arctic sea-ice cover: Insights from remote sensing observations and model reanalysis. *Heliyon* **2020**, *6*, e04355. [CrossRef]
30. Proshutinsky, A.; Krishfield, R.; Toole, J.M.; Timmermans, M.; Williams, W.; Zimmermann, S.; Yamamoto-Kawai, M.; Armitage, T.W.K.; Dukhovskoy, D.; Golubeva, E.; et al. Analysis of the Beaufort Gyre Freshwater Content in 2003–2018. *J. Geophys. Res. Ocean.* **2019**, *124*, 9658–9689. [CrossRef]
31. Li, C. 3D analytic model for testing numerical tidal models. *J. Hydraul. Eng.* **2001**, *127*, 709–717. [CrossRef]
32. Li, C. Can friction coefficient be estimated from cross stream flow structure in tidal channels? *Geophys. Res. Lett.* **2003**, *30*, 1987. [CrossRef]

33. Li, C.; Boswell, K.; Chaichitehrani, N.; Huang, W.; Wu, R. Weather induced subtidal flows through multiple inlets of an arctic microtidal lagoon. *Acta Oceanol. Sin.* **2019**, *38*, 1–16. [CrossRef]
34. Li, C. Tidally Induced Residual Circulation in Estuaries with Cross Channel Bathymetry. Ph.D. Thesis, University of Connecticut, Storrs, CT, USA, 1996; 242p.
35. Parker, B.B. Frictional Effects on the Tidal Dynamics of Shallow Estuary. Ph.D. Thesis, The Johns Hopkins University, Baltimore, MA, USA, 1984; 291p.
36. Proudman, J. *Dynamical Oceanography*; Methuen: London, UK, 1953; 409p.
37. Butterworth, S. On the theory of filter amplifiers. *Exp. Wirel. Wirel. Eng.* **1930**, *7*, 536–541.
38. Li, C. Subtidal Water Flux through a Multi-inlet System: Observations Before and During a Cold Front Event and Numerical Experiments. *JGR-Ocean.* **2013**, *118*, 1–16. [CrossRef]
39. Li, C.; Weeks, E.; Huang, W.; Milan, B.; Wu, R. Weather-Induced Transport through a Tidal Channel Calibrated by an Unmanned Boat. *J. Atmos. Ocean. Technol.* **2018**, *35*, 261–279. [CrossRef]
40. Li, C.; Armstrong, S.; Williams, D. Residual Eddies in a Tidal Channel. *Estuaries Coasts* **2006**, *29*, 147–158. [CrossRef]

sensors

MDPI

Article

# Correction of Biogeochemical-Argo Radiometry for Sensor Temperature-Dependence and Drift: Protocols for a Delayed-Mode Quality Control

Quentin Jutard [1], Emanuele Organelli [2,*], Nathan Briggs [3], Xiaogang Xing [4], Catherine Schmechtig [1], Emmanuel Boss [5], Antoine Poteau [6], Edouard Leymarie [6], Marin Cornec [6], Fabrizio D'Ortenzio [6] and Hervé Claustre [6]

1 CNRS & Sorbonne Université, OSU Ecce Terra, 4 Place Jussieu, CEDEX 05, 75252 Paris, France; quentin.jutard@sorbonne-universite.fr (Q.J.); catherine.schmechtig@imev-mer.fr (C.S.)
2 National Research Council (CNR), Institute of Marine Sciences (ISMAR), Via del Fosso del Cavaliere 100, 00133 Rome, Italy
3 National Oceanography Centre, Southampton SO14 3ZH, UK; nathan.briggs@noc.ac.uk
4 State Key Laboratory of Satellite Ocean Environment Dynamics, Second Institute of Oceanography, Ministry of Natural Resources, 36 Baochubei Road, Hangzhou 310012, China; xing@sio.org.cn
5 School of Marine Sciences, University of Maine, Orono, ME 04469, USA; emmanuel.boss@maine.edu
6 CNRS & Sorbonne Université, Laboratoire d'Océanographie de Villefranche, 06230 Villefranche sur mer, France; antoine.poteau@imev-mer.fr (A.P.); edouard.leymarie@imev-mer.fr (E.L.); marin.cornec@imev-mer.fr (M.C.); fabrizio.dortenzio@imev-mer.fr (F.D.); herve.claustre@imev-mer.fr (H.C.)
* Correspondence: emanuele.organelli@cnr.it; Tel.: +39-06-45488289

**Abstract:** Measuring the underwater light field is a key mission of the international Biogeochemical-Argo program. Since 2012, 0–250 dbar profiles of downwelling irradiance at 380, 412 and 490 nm besides photosynthetically available radiation (PAR) have been acquired across the globe every 1 to 10 days. The resulting unprecedented amount of radiometric data has been previously quality-controlled for real-time distribution and ocean optics applications, yet some issues affecting the accuracy of measurements at depth have been identified such as changes in sensor dark responsiveness to ambient temperature, with time and according to the material used to build the instrument components. Here, we propose a quality-control procedure to solve these sensor issues to make Argo radiometry data available for delayed-mode distribution, with associated error estimation. The presented protocol requires the acquisition of ancillary radiometric measurements at the 1000 dbar parking depth and night-time profiles. A test on >10,000 profiles from across the world revealed a quality-control success rate >90% for each band. The procedure shows similar performance in re-qualifying low radiometry values across diverse oceanic regions. We finally recommend, for future deployments, acquiring daily 1000 dbar measurements and one night profile per year, preferably during moonless nights and when the temperature range between the surface and 1000 dbar is the largest.

**Keywords:** BGC-Argo; radiometry; quality control

## 1. Introduction

The international Biogeochemical-Argo (i.e., BGC-Argo) program has revolutionized the way we acquire measurements of biogeochemically relevant variables in the open ocean [1,2]. In 2016, the Biogeochemical-Argo planning group has defined six core variables to accomplish the scientific and observational objectives of the program that include the study of the ocean carbon uptake and acidification, oxygen minimum zones and nitrate cycling, biological carbon pump, phytoplankton communities, and joint use with ocean color satellite observations [3]. In particular, to study phytoplankton dynamics and combine in-situ with remote sensing observations, radiometry, i.e., measurements of downwelling irradiance ($E_d$) and photosynthetically available radiation (PAR), has been selected as a core variable.

Currently, the BGC-Argo program has accumulated more than 40,000 profiles of downwelling irradiance (between 0 and 250 dbar), acquired by more than 100 floats in the global ocean, across a variety of trophic and environmental conditions, and in remote regions (https://biogeochemical-argo.org/, accessed on 13 September 2021). These profiles have proved to be fruitful measurements for diverse applications. Downwelling irradiances at various wavelengths have been implied in the analysis of the bio-optical behavior of the global ocean [4] and the dynamics of dissolved organic matter [5,6], and for the validation of space-based ocean color measurements and products [7–13]. Besides, $E_d$ and PAR have been widely used to understand particulate organic carbon fluxes and export [14–16], to study phytoplankton dynamics [17–25], and to improve numerical and radiative-transfer models [26,27].

Despite the relevant scientific results, some inconsistencies in deep radiometric measurements, where the lowest irradiances are expected, have been observed [8,9,28]. With time and through the analysis of acquired data, our knowledge on the sensor behavior has progressively improved and identified two main radiometer characteristics which are independent one from the other, neglected since the launch of the fleet in 2012. First, the dark measurements of the sensors are sensitive to the ambient temperature which ultimately reduces measurement accuracy, especially in the deep part of the profile where the remaining light is very low [8,9]. Such variance in the sensor responsivity with environmental temperature is radiometer component- and wavelength-dependent [29]. Indeed, we have observed that the sensor dark dependence on temperature is conditioned by the material used to build the sensor container, i.e., aluminum or polyether-ether-ketone (hereafter PEEK). Laboratory experiments have confirmed this temperature dependence for radiometers to be deployed in Arctic waters [30] and showed differences between those made in aluminum and PEEK across a wide range of ambient temperatures (see Supplementary Materials Section S1). Second, the sensors' dark measurements may drift after several years of float operation. Radiometers mounted on Argo floats have not been equipped with mechanical shutters that acquire along cast dark measurements during daylight profiles, mainly due to relevant power consumption. We thus evolved the initially established sampling protocol towards the acquisition of reference night profiles and dark measurements at the 1000 dbar parking depth over the whole float lifetime in order to characterize, quality-control and solve these sources of variability in the sensor response.

As for all Argo physical and biogeochemical variables, radiometry quality-control (QC) must be provided in real-time (RT) and delayed-mode (DM). The RT-QC is mainly devoted to operational oceanography (e.g., assimilation in forecast models of ocean state) and consists in a number of automatic procedures that target the evaluation of a single profile at a time and QC data distribution within 12 h from sampling. The DM-QC aims to make data available within 12 months from the acquisition, after human control and exploiting all measured profiles together [31]. The resulting DM-QC dataset is expected to have the highest quality requested for scientific analysis and, ultimately, for climate studies. The RT-QC procedure for radiometry, accepted by the Argo Data Management Team, aims to check and flag measurements outside the range of expected values [32]. Alternatively, Organelli et al. [28] have proposed a near-real-time methodology detecting environmental signals in radiometric profiles due to clouds and wave focusing near surface, that is dedicated to bio-optical and remote-sensing applications (i.e., calm sea and uniform sky conditions during the measurement [33]). No DM-QC for radiometric data, as well as methods to characterize and solve sensor dark dependency on temperature and drift have been implemented yet.

Here, we will exploit the global array of floats equipped with radiometers to develop and assess a DM-QC procedure that aims to correct the effect of changes in environmental temperature on BGC-Argo radiometric dark signals according to the material used to build the instrument, and account for sensor dark drift with time (hereinafter referred to as aging).

Following Equation (1) we convert digital counts (DC) to irradiance (units of W $m^{-2}$ $nm^{-1}$) and PAR (units of μmol photons $m^{-2}$ $s^{-1}$) values:

$$E_d(\lambda) = Im(\lambda) * a_1(\lambda, T_s, t) * (DC(\lambda) - a_0(\lambda, T_s, t)) \quad (1)$$

this study will focus on the correction of the effects of the time, $t$, and of the sensor internal temperature, $T_s$, on the $a_0$ calibration coefficient for each band of each sensor (i.e., the dark signal), as the temperature dependency of the calibration coefficient $a_1$ has been found to be negligible [34]. *Im* is the immersion coefficient, fixed for each band. We will discuss procedure performance and show examples for a variety of trophic and illumination conditions encountered across the global ocean. Finally, we will present advantages, limitations and recommendations for the method. We anticipate the proposed methodology and the recommended sampling protocol will open the door to the operational distribution of the highest quality Argo radiometric profiles to the international oceanographic community. All symbols and abbreviations used here are listed in the nomenclature list given below.

## 2. Materials and Methods

### 2.1. The Biogeochemical-Argo Database

Biogeochemical-Argo data used to develop and assess the DM-QC procedure for radiometric profiles were acquired by 55 no longer profiling PROVOR-CTS4 floats, for a total of 12,867 measured radiometry profiles. This fleet has operated since 2012 across a variety of trophic environments and regional seas (Figure 1). All floats were configured and deployed according to standard procedures [35]. The data were downloaded from the Coriolis Global Data Assembly Center (GDAC) and stored in the Argo B and trajectory files (ftp://ftp.ifremer.fr/ifremer/argo (accessed on 1 November 2020)).

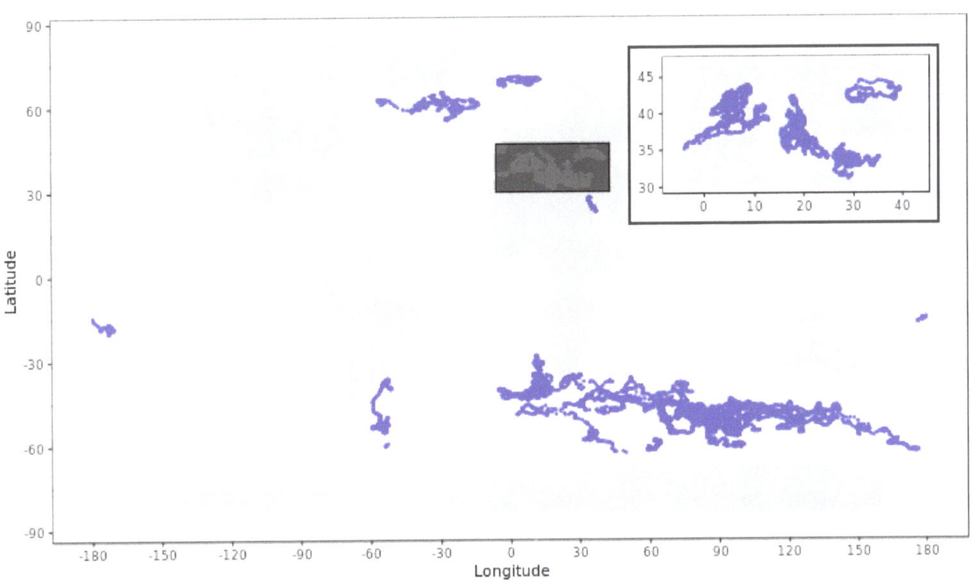


**Figure 1.** Sampled stations by the 55 profiling BGC-Argo floats considered in this study.

Floats were programmed to drift at a parking depth of 1000 dbar and acquire vertical profiles up to the sea surface every 1 to 10 days. Pressure and water temperature data were collected every 2 s by a SBE-41 CP conductivity-temperature-depth sensor (Sea-Bird Scientific, Bellevue, WA, USA), and quality-controlled according to standard, internationally-accepted protocols [36]. $E_d$ at three wavelengths (380, 412 and 490 nm) and PAR measurements were acquired by an OCR-504 radiometer (Sea-Bird Scientific), without an internal temperature probe and configured with a different sensor for each channel [37]. Though all the floats were equipped with the same radiometer model, the thermodynamic properties of five instruments made with aluminum (i.e., 693 profiles), deployed between 2014 and 2018, were different from those made with PEEK (see Supplementary Materials Section S1).

Radiometric profiles were acquired in the upper 250 dbar, around local noon to reduce the impact of low solar zenith angles [33]. To develop specific correction procedures for the dark correction, which is known to be temperature-dependent [38,39], night profiles (i.e., solar elevation < 5°) were acquired across a similar temperature range as day profiles since 2014, but neither systematically nor homogeneously among all floats. Moreover, radiometric measurements were also acquired daily during the float drift at the 1000 dbar parking depth to evaluate any change in the instrument's response with time. This was implemented mid-2014 for all floats but those deployed in the Baffin Bay (Arctic Sea). Hereafter, we will refer to radiometric data used to develop and assess the DM-QC control such as: (i) day profiles (high light and high temperature variability); (ii) night profiles (no or very dim light with high temperature variability); and (iii) drift measurements (no light and low temperature variability).

In the following sections, we will show that both the acquisition of night profiles and daily radiometric measurement at 1000 dbar represent key ancillary measurements to correct the sensor's dark signal and develop the most accurate DM-QC procedure. However, since in the Coriolis GDAC there are additional 11350 profiles acquired by 76 no longer profiling floats without sufficient ancillary night profiles or drift measurements acquired for longer than 80% of the float lifetime (Table 1), we have developed specific DM-QC procedures for those floats that are presented in the Supplementary Material Section S2. Hence, the following sections will only focus on the best possible DM-QC method that we recommend for future BGC-Argo radiometry deployments.

**Table 1.** Availability of night profiles and daily drift measurements for the 55 and 76 BGC-Argo floats.

| OCR 504 Model | Drift Acquired for > 80% of the Float Lifetime | | Drift Acquired for ≤ 80% of the Float Lifetime | | Total |
|---|---|---|---|---|---|
| | Night Profiles | No Night | Night Profiles | No Night | |
| PEEK | 50 | 10 | 32 | 17 | 109 |
| Aluminum | 5 | 1 | 9 | 7 | 22 |
| All | 55 | 11 | 41 | 24 | 131 |

*2.2. Reconstruction of the Sensor Internal Temperature*

The thermodynamics response of the sensor is not instantaneous (see Supplementary Materials Section S1), thus the radiometer internal temperature must be reconstructed to develop the DM-QC procedure. Following laboratory experiments (see Supplementary Materials Section S1), the internal temperature $T_s$ at which the sensor operates was modeled using a delay first-order differential equation:

$$\frac{1}{k}\frac{dT_s}{dt}(t) = T_w(t - \Delta t) - T_s(t) \quad (2)$$

where $T_w$ is the temperature of the surrounding water; $k$ and $\Delta t$ are empirically estimated coefficients which represent the physical characteristics of the radiometer (Table 2).

**Table 2.** Parameters used to reconstruct the sensor internal temperature $T_s$ according to the material of the radiometer components.

| OCR 504 Model | k | $\Delta t$ |
|---|---|---|
| PEEK | 0.2 $min^{-1}$ | 1 min |
| Aluminum | 0.44 $min^{-1}$ | 0.25 min |

To integrate Equation (2) along the entire profile, the following assumptions were made:

1) $T_s = T_w$ at the bottom of the profile. All floats spend at least one day at 1000 dbar before profiling. Thus, when the float starts acquiring measurements, the sensor temperature is at the equilibrium with the environment ($1/k + \Delta t << 1$ day);
2) The ascending speed of the float, $c$, is assumed to be constant, thus $c$ = 0.1 dbar $s^{-1}$. We analyzed 27,000 profiles from 165 PROVOR CTS-4 Argo floats, and found that 91% of the profiles showed an average ascending speed ranging between 0.08 dbar $s^{-1}$ and 0.12 dbar $s^{-1}$ (Figure 2). A sensitivity test on correction of $E_d(490)$ for the float WMO 6901654 revealed that, when using 0.08 and 0.12 dbar $s^{-1}$ instead of 0.1 dbar $s^{-1}$, the corrected $E_d(490)$ values change by at most $1.7 \times 10^{-5}$ W $m^{-2}$ $nm^{-1}$, with 95% of the measurement points vary by less than $5.3 \times 10^{-6}$ W $m^{-2}$ $nm^{-1}$. This observed variability is consistent with the manufacturer-established sensor noise of $2.5 \times 10^{-5}$ W $m^{-2}$ $nm^{-1}$ [37].

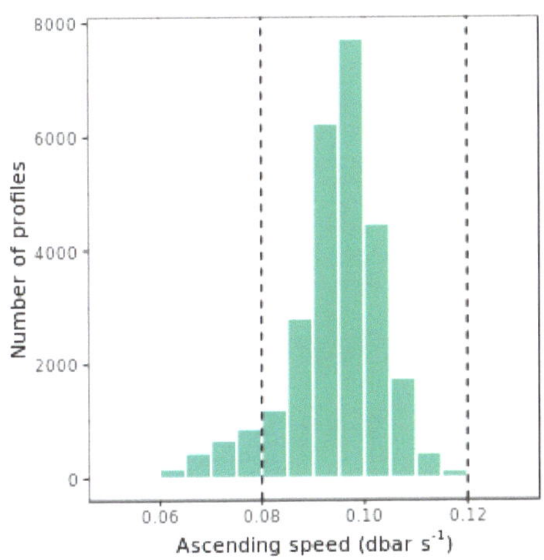


**Figure 2.** Histogram of the average float ascent speed for 27000 BGC-Argo radiometry profiles, reconstructed from the available time stamps in the trajectory profile. Vertical dashed lines indicate the two values used for the sensitivity test which interval includes 91% of tested profiles.

We then introduce $T_s^*$ which is $T_s$ delayed by $\Delta t$. This allows Equation (2) to be rewritten as an ordinary differential equation:

$$\frac{1}{k}\frac{dT_s^*}{dt} = T_w - T_s^* \tag{3}$$

with:

$$T_s^* = T_s(t + \Delta t) \tag{4}$$

Temperature is measured along a discrete axis of corresponding pressure measurements. We numerically integrate Equation (3) along this discrete axis with index 0 corresponding to the deepest (and first) measurement. We also introduce $t_n$, i.e., the time at which each measurement is taken, with $t_0 = 0$, and $P_{w_n}$ which is the pressure measurement associated to $T_{w_n}$.

From Assumption 1 described above:

$$T^*_{s_0} = T_{s_0} = T_{w_0} \tag{5}$$

Equation (3) can be discretized as:

$$T^*_{s_n} = T^*_{s_{n-1}} + k * (t_n - t_{n-1}) * \left(T_{w_{n-1}} - T^*_{s_{n-1}}\right) \tag{6}$$

Using Assumption 2, we can express:

$$t_n = c^{-1} * (P_{w_0} - P_{w_n}) \tag{7}$$

so that Equation (6) becomes:

$$T^*_{s_n} = T^*_{s_{n-1}} + \frac{k}{c} * (P_{w_{n-1}} - P_{w_n}) * \left(T_{w_{n-1}} - T^*_{s_{n-1}}\right) \tag{8}$$

Equation (8) can be computed to obtain $T^*_{s_n}$ for each $P_{w_n}$ value. The pressure axis $P_{s_n}$ is then defined as:

$$P_{s_n} = P_{w_n} + c * \Delta t \tag{9}$$

so that for each $n$, $T_{s_n}$ is equal to $T^*_{s_n}$ when $T_{s_n}$ values are associated to the pressure axis $P_{s_n}$.

The final step is to interpolate $T_{s_n}$ to retrieve $T_s$ values that correspond to the pressure axis of radiometric measurements.

To reconstruct the sensor internal temperature for radiometric measurements acquired during the float drift at the 1000 dbar parking depth, the model described by Equations (2)–(9) could not be applied because of the low frequency of drift measurements and the inapplicability of Assumption 2. In this case, because water temperature changes slowly during the drift of the float, and the float spends at least one day at those given depth and temperature, the closest (in time) water temperature measurement to the radiometry sampling was selected as the corresponding $T_s$.

## 3. Protocols for the Correction of Aging and Temperature Dependence of the Dark Signal

### 3.1. Theoretical Framework

The measured irradiance $E_{d_{meas}}$ is described as a function of the real irradiance $E_{d_{real}}$, the sensor internal temperature $T_s$, the time $t$, and the sensor random normal noise $\varepsilon$:

$$E_{d_{meas}} = F\left(E_{d_{real}}, T_s, t\right) + \varepsilon \tag{10}$$

We assumed that:

$$E_{d_{meas}} = h(T_s, t) * E_{d_{real}} + f(T_s) + g(t) + \varepsilon \tag{11}$$

where $h$ is the slope error introduced by the temperature effects and aging, $f(T_s)$ and $g(t)$ are the dark errors introduced by the sensor temperature and aging respectively, which are assumed to be independent from one another.

For night profiles and drift measurements, the float is in the dark so that $E_{d_{real}}$ is assumed equal to 0. Equation (11) is thus modified to:

$$\text{For night profiles, } E_{d_{meas}} = 0 + f(T_s) + g(t) + \varepsilon \tag{12}$$

$$\text{For drift measurements, } E_{d_{meas}} = 0 + f(T_s \sim constant) + g(t) + \varepsilon \tag{13}$$

In Equation (13), we indicate that the water temperature variations at the 1000 dbar parking depth are relatively small, which means $T_s$ can be considered as near constant. This also means that drift measurements at 1000 dbar parking depth can be used to estimate the sensor's dark aging $g(t)$ almost independently from changes in the environmental temperature. This estimated $g(t)$ is then needed in Equation (12) to estimate the sensor's dark temperature dependency $f(T_s)$ using night profiles, which are acquired over a larger range of temperatures than drift measurements. This is the rationale to estimate $g(t)$ and perform the correction for sensor dark aging before the estimation of $f(T_s)$ and the correction of the sensor temperature-dependence.

### 3.2. Overview of the Procedure

The overall quality-control procedure includes five consecutive steps, which will be described in the following sections, and are the same both for $E_d(\lambda)$ and PAR: (i) Visual quality control; (ii) Correction of the sensor aging; (iii) Correction of the sensor temperature-dependence; (iv) Error estimation; and (v) Assignment of quality flags.

The overview of the whole procedure to correct for aging and then temperature-dependence of the dark sensor is shown in Figure 3. After the visual check, the workflow starts with the computation of a multiple linear or linear-quadratic regression that must be visually checked by the DM operator before applying the aging correction to all measured profiles of a given float. We remind that, for BGC-Argo DM-QC, the operator must use own scientific expertise and provide critical inputs to evaluate the correction results. If the correction for the aging does not yield satisfactory results, the DM operator may move to the following step. This is recommended for floats with short lifespan.

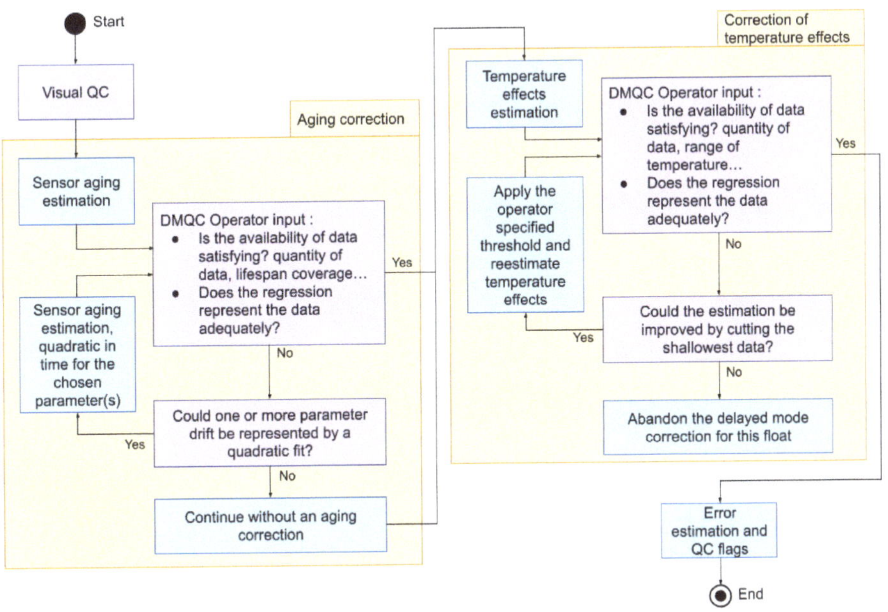


**Figure 3.** Flowchart of the QC procedure to correct radiometry for aging and temperature dependency.

Corrected profiles are then adjusted for the temperature-dependence by computing linear regressions on night profiles. The linear regression must be visually checked by the DM operator before applying the correction to all measured profiles of a given float. The DM operator must thus evaluate that the temperature range covered by night profiles is representative of the temperature variability encountered by the float over the whole lifetime, as well as the regression fit to the data. If the method does not yield satisfactory results, the DM operator abandons the quality control of that float. An example of unsatisfactory linear regression is shown in Supplementary Material Section S3. If the correction is successful the error associated to each measurement is estimated and quality flags are assigned.

### 3.2.1. Visual Quality Control

According to the standard Argo procedures [36], the DM-QC includes a preliminary visual check, profile by profile, made by the operator before the application of automatic correction routines. Thus, each data point within the profile is ultimately assigned one of the standard Argo QC flags: "1" for good data, "2" for probably good data; "3" for probably bad data; and "4" for bad data. Both flags 1 and 2 will be used to correct sensor's dark aging and temperature dependence as described here below.

Practically, the visual check starts from the evaluation of RT-QC radiometry data [32]. The DM operator first evaluates if RT-QC Flag "3" measurements must be confirmed as bad or upgraded to "1" or "2". Then, the operator visually detects any obvious outlier along the profile which is not related to environmental signals due to clouds and wave focusing/defocusing. The outliers are assigned to Flag "3" and "4" depending on the DM operator's confidence. Radiometric measurements flagged as "3" and "4" are not further evaluated and are excluded from the following QC steps.

### 3.2.2. Correction of the Sensor Dark's Aging

In the following section, the protocol to correct the sensor dark's aging which is based on the use of drift measurements is presented. Outliers are first removed from drift measurements and are defined as any value falling outside of the range between [1st_quartile–1.5*(3rd_quartile–1st_quartile)] and [3rd_quartile + 1.5*(3rd_quartile–1st_quartile)].

Following Equation (13), $E_{d_{meas}}$ is equal to 0 and $T_s$ at 1000 dbar shows relatively low variance. However, this small variance can still have a visible impact on the drift data (Figure 4). Apart from deviations due to temperature, the sensor aging most often appears as a linear function of time. Thus, $g(t)$ is estimated by applying a multiple linear regression model of $E_{d_{meas}}$ as a function of $t$ and $T_s$:

$$E^*_{d_{meas}} = Ad + Bd * T_s + Cd * t \tag{14}$$

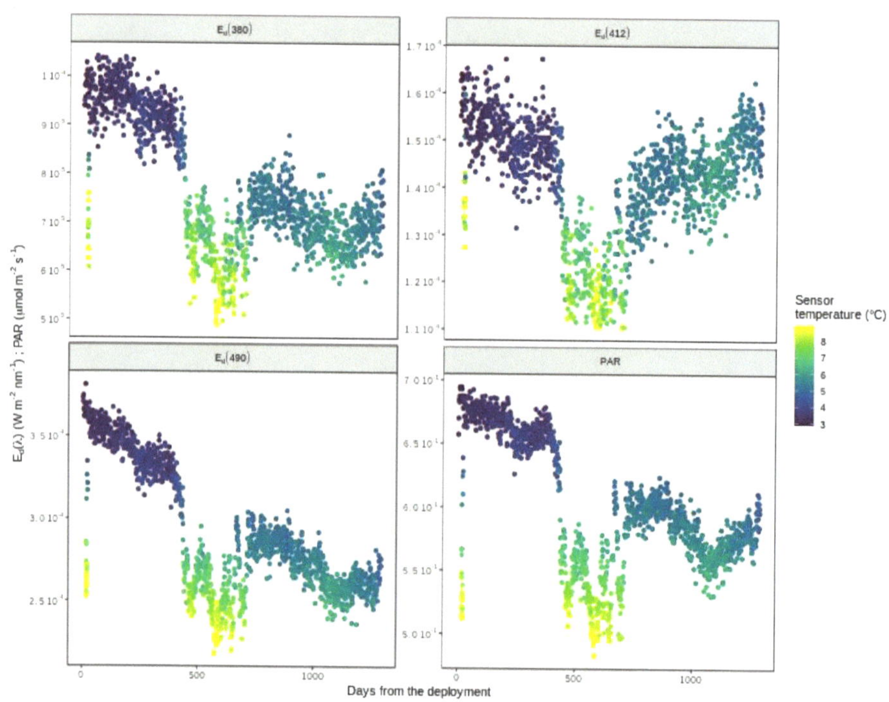


**Figure 4.** Radiometry drift measurements for $E_d(\lambda)$ and PAR as a function of time and temperature. Example is shown for the float WMO6901584.

Subsequently, the DM operator must visually check the resulting fit from Equation (14) by estimating $E_d$ at a reference temperature which has been set to 5 °C (Figure 5):

$$E_{d_{5c}} = E_{d_{meas}} - Bd * (T_s - 5) \tag{15}$$

and:

$$E^*_{d_{5C}} = Ad + 5Bd + Cd * t \tag{16}$$

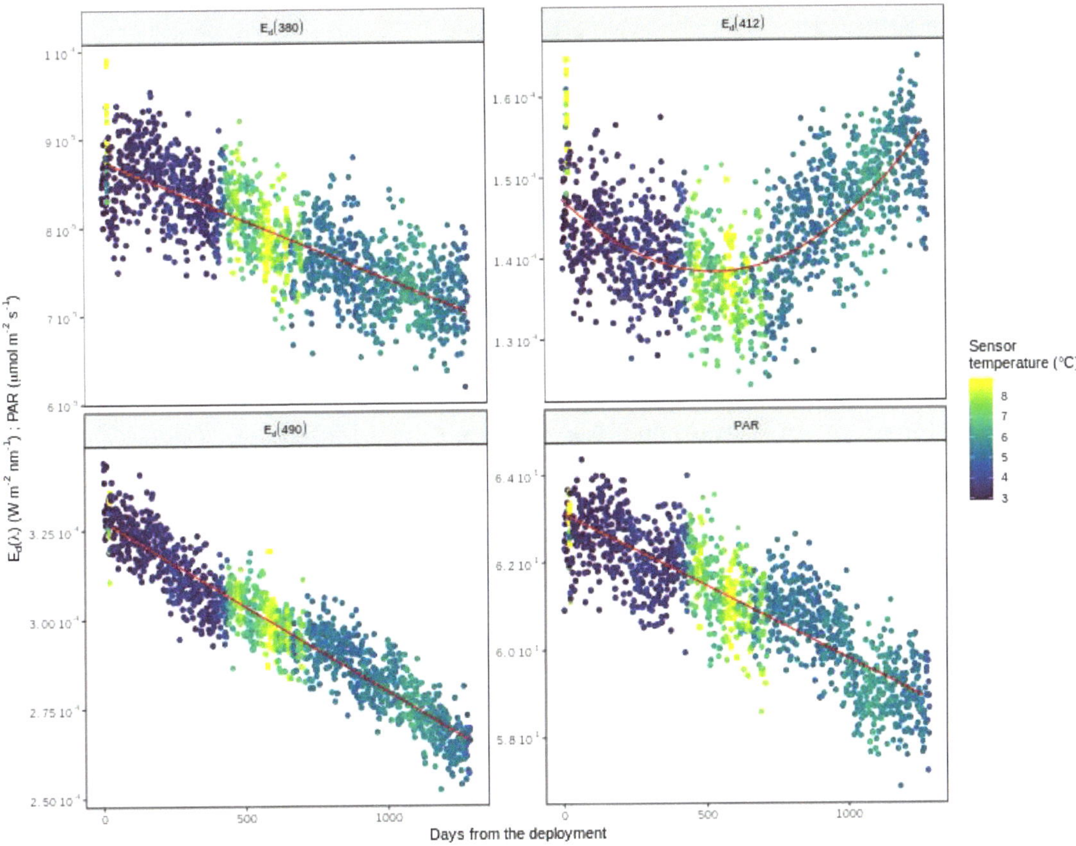


**Figure 5.** Radiometry drift measurements for $E_d(\lambda)$ and PAR as a function of time after estimation at a reference temperature of 5 °C. Solid line is the fit to all points. For this float, the fit is linear for all channels but $E_d(412)$. Example is shown for the float WMO6901584.

Because the aging may change sign and/or intensity over time (e.g., $E_d(412)$ in Figure 4), the DM-QC operator may not be satisfied with the results of the linear fit in Equation (14). In such a case, the operator may decide to fit $E_{d_{meas}}$ by a quadratic function versus $t$ and linear versus $T_s$ ($E_d(412)$ in Figure 5):

$$E^*_{d_{meas}} = Ad + Bd * T_s + Cd * t + Qd * t^2 \tag{17}$$

so that:

$$g(t) = Ad + Cd * t + Qd * t^2 \tag{18}$$

where $Qd$ is 0 when the linear regression in Equation (14) is applied. It should be noted that Equation (18) includes the constant offset $Ad$ from the bilinear regression.

*Ad* is not mathematically required to compute $g(t)$ because another coefficient will be computed when temperature correction is performed (see following sections). However, it is here included in order to allow the DM operator to run the procedure using realistic radiometric values.

The multiple linear model described by Equation (14) is able to correct for the small temperature variations found at the 1000 dbar parking depth. However, this temperature correction cannot be applied to the whole profiles because they span a large range of variability in temperature so that estimated coefficients from Equation (14) are not suitable. In addition, the estimation at a reference temperature of 5 °C allows the DM operator to visualize and evaluate, float by float, the goodness of the aging's correction procedure. However, if the operator is still not satisfied with the proposed correction after visual check, we suggest to proceed with the temperature-dependence correction anyway and test the results. This is especially recommended for floats with a short lifespan.

#### 3.2.3. Correction of the Sensor Dark's Temperature Dependence

In this section, the protocol to correct the sensor dark's dependence on temperature which is based on the use of night profiles is presented. We recall that $E_{d_{real}}$ is assumed equal to 0 along the whole night profile, which covers a large variability in water temperature. As a first step, all night profiles collected by a single float are corrected for the sensor aging as described above. $E_{d_{night}}$ is then defined as:

$$E_{d_{night}} = E_{d_{meas}} - g(t) = E_{d_{meas}} - Ad - Cd * t - Qd * t^2 = f(T_s) + \varepsilon \quad (19)$$

Then, $E_{d_{night}}$ is linearly fitted as a function of the reconstructed sensor internal temperature $T_s$:

$$E^*_{d\,night} = At + Bt * T_s \quad (20)$$

Figure 6 shows an example of aging-corrected night profiles and regression analysis. It is important to note that some night profiles might be influenced by the moon and star light or acquired close to dawn and dusk. To remove such polluted data, the DM operator may select a pressure threshold.

Subsequently, the offset to correct for sensor darks' dependence on temperature is expressed as:

$$f(T_s) = At + Bt * T_s \quad (21)$$

The final correction to be applied to all 0–250 dbar profiles is finally expressed as:

$$E_{d_{corr}} = E_{d_{meas}} - f(T_s) - g(t) \quad (22)$$

$$E_{d_{corr}} = E_{d_{meas}} - At - Bt * T_s - Ad - Cd * \mathrm{t} - Qd * t^2 \quad (23)$$

$$E_{d_{corr}} = E_{d_{meas}} - A - B * T_s - \mathrm{C} * \mathrm{t} - Q{*}t^2 \ (24) \quad (24)$$

where $A = At + Ad$, $B = Bt$, $C = Cd$, and $Q = Qd$. It must be noted that the corrected irradiance $E_{d_{corr}}$ is not equal to $E_{d_{real}}$ (Equation (11)) as only the temperature and aging effects on the dark signal have been corrected. To equate $E_{d_{corr}}$ and $E_{d_{real}}$, $h(T_s,t)$ in Equation (11) must be assumed equal to 1.

#### 3.2.4. Error Estimation

Upon implementation of corrections presented above, the error associated with each measured value ($\sigma_{Ed}$) is estimated as the maximum value between the Noise Equivalent Irradiance (NEI) (as provided by the manufacturer), and the relative error (ER) multiplied by the corrected radiometry value $E_{d_{corr}}$:

$$\sigma_{E_d} = \max\left(NEI_{E_d}; ER_{E_d} * E_{d_{corr}}\right) \quad (25)$$

$NEI_{E_d}$ is the manufacturer's NEI value of OCR-504 radiometers equal to $2.5 \times 10^{-5}$ W $m^{-2}$ $nm^{-1}$ for all $E_d(\lambda)$ [37]. For PAR, $NEI_{E_d}$ was estimated by computing the maximum standard deviation observed for the dark values at the 1000 dbar parking depth corrected for any aging among a total of 34 selected floats. The resulting $NEI_{E_d}$ for PAR is equal to 0.03 µmol photons $m^{-2}$ $s^{-1}$. ER is 5% for PAR [40] and 2% for $E_d(\lambda)$ following previous calibration error estimations [41,42].

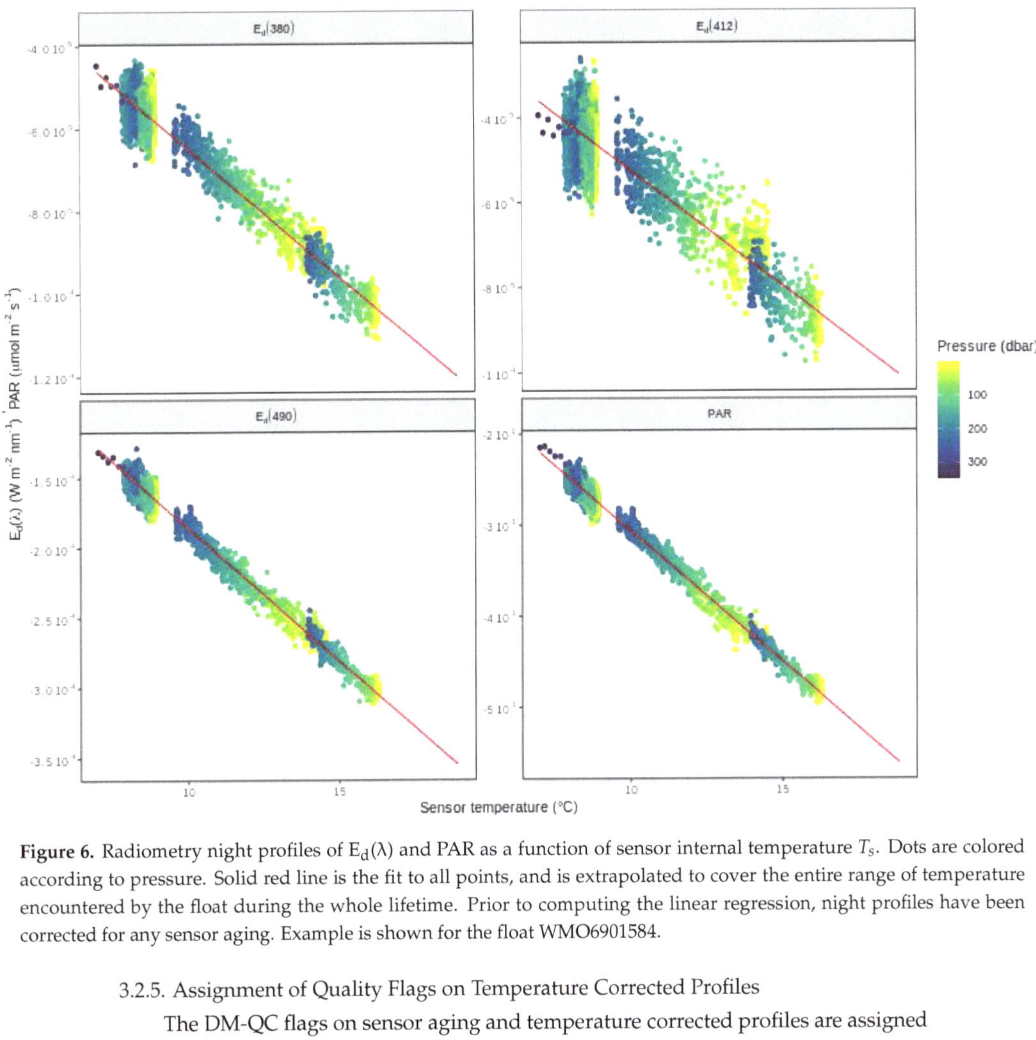


**Figure 6.** Radiometry night profiles of $E_d(\lambda)$ and PAR as a function of sensor internal temperature $T_S$. Dots are colored according to pressure. Solid red line is the fit to all points, and is extrapolated to cover the entire range of temperature encountered by the float during the whole lifetime. Prior to computing the linear regression, night profiles have been corrected for any sensor aging. Example is shown for the float WMO6901584.

### 3.2.5. Assignment of Quality Flags on Temperature Corrected Profiles

The DM-QC flags on sensor aging and temperature corrected profiles are assigned according to the following procedure:

- Recover the QC flags assigned with the visual QC. These profiles contain Flags "1", "2", "3" and "4";
- Detect the dark values within corrected profiles applying successive Lilliefors tests ($\alpha = 0.01$; ref. [28]), and assign Flag "2";
- Change radiometry flags "3" or "4" due to visual QC to "4";
- If pressure QC flag is "3" or "4", radiometry flag is assigned as "4";
- If $T_S$ cannot be reconstructed, the radiometry flag is assigned as "4".

## 4. Performance of the DM-QC Procedure

The DM-QC procedure described above to correct for sensor dark changes with time and varying environmental temperature was tested over a total of 55 BGC-Argo profiling floats with ancillary night profiles and drift measurements acquired over more than 80% of the float lifetime. All these floats, operating across the globe, were equipped with OCR-504 radiometers and acquired 0–250 dbar $E_d$ profiles at 380, 412 and 490 nm in addition to PAR.

In Figure 7, we show examples of vertical profiles before and after correction for sensor's dark aging and temperature dependence. The magnitude of the correction applied as represented by the A, B, and C parameters obtained through Equation (24), and its variability over the ensemble of floats whose sensor aging was corrected linearly are shown for each band in Figure S9 (Supplementary Materials Section S4). The distributions of the A, B, and C parameters were generally normal and, the impact of temperature on the sensor's dark signal showed to be larger than the one due to the sensor's aging.

Examples of corrected profiles (Figure 7) encompass a variety of oceanic environments with diverse optical, trophic and biogeochemical conditions [4,20,43,44], thus showing applicability of the procedure at the global scale. In particular, the steps we set up for the DM-QC BGC-Argo radiometry (Figure 3) provide adjustments of specific features that characterize the profiles (Figure 7). First of all, all non-zero dark measurements at depth are shifted to zero or re-qualified as very low irradiance measurements that, otherwise, would have been disregarded. Indeed, the DM-QC procedure makes vertical profiles usable at greater depths so that biogeochemical, modelling, and optical applications can be enhanced. This is particularly relevant for permanently oligotrophic clear waters (e.g., mid-ocean gyres; Figure 7d) where sunlight around local noon can penetrate deeper than 250 dbar [8], or in productive high-latitude seas during wintertime where the underwater light field can expand down to 150 dbar (Figure 7j) and contribute to phytoplankton blooms [18].

Contrarily, in the upper part of the profile where irradiance values are the highest and aging and temperature issues are expected to have a negligible impact [8,28], the developed correction protocols do not determine significant changes in the measured values (Figure 7). In addition, the developed QC procedure does not affect the signature of the environmental signals such as those due to clouds and wave focusing/defocusing (Figure 7a). Such characteristics reinforce previously published scientific studies restricted to the first optical depth or the mixed layer [4,6,20], and joint applications with remote sensing observations [8,11]. Yet a newly generated radiometric database enhanced with sensor dark's aging and temperature-dependence corrections will surely open to the possibility of re-analysis studies.

However, the applied DM procedure correctly resolves artificial features such as steps in the profiles due to a significant increase of the dark counts which respond to the sudden changes in water temperature (Figure 7g–i). The developed protocols remove these features and shift to zero dark values at depth, so that the resulting radiometric profiles show the monotonic decrease with depth as expected.

The DM-QC procedure we developed has been implemented over a total of 12,867 measured profiles each band. The procedure returned profiles that monotonically decreased as expected from theory and reached greater depths (Figure 7). A total of 11,824 profiles (from 47 floats), i.e., about 92% of the tested database for bands at 412 and 490 nm, and PAR was corrected (Figure 8). In the case of $E_d(380)$, correction was successful for 11,597 profiles (from 46 floats), i.e., 90% of the tested database. In particular, the DM successfully corrected profiles derived from 45 floats made with PEEK components (44 floats for $E_d(380)$), and two floats with aluminum components. The uncorrected 227 $E_d(380)$ profiles (all from one float) were corrected with alternative procedures (see Supplementary Materials Section S2). The developed QC procedure demonstrated high and similar performances for all radiometric channels. This suggests strong potential to implement these DM-QC protocols to other wavelengths and, ultimately to hyperspectral radiometers.

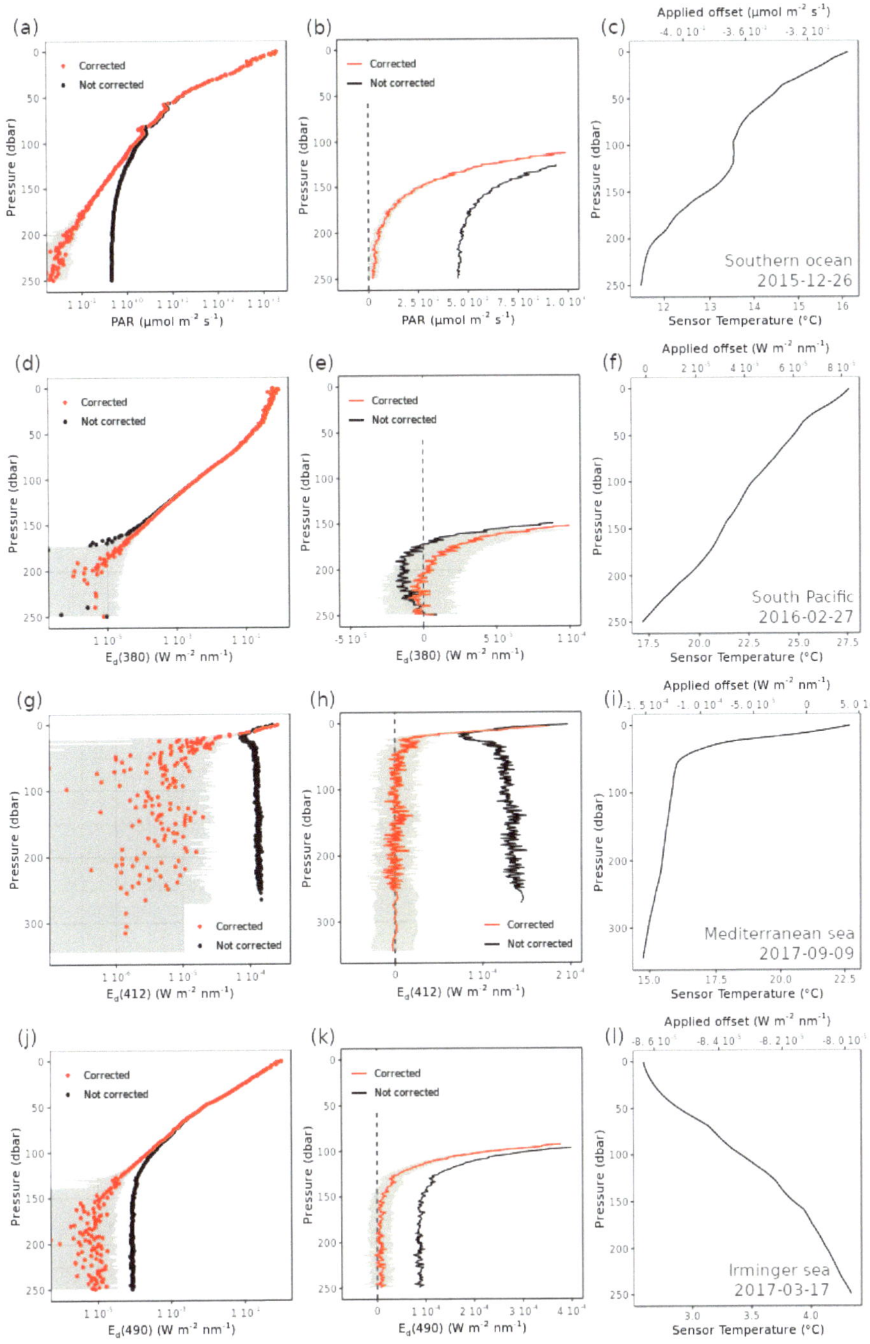


**Figure 7.** Examples of radiometry profiles before and after DM-QC: Left) profiles are shown in a semi-log scale; Centre) profiles are shown in a linear scale; Right) the reconstructed sensor internal temperature $T_s$ is shown (Equations (2)–(9)). Examples derive from four BGC-Argo floats deployed in oceanic regions characterized by diverse trophic and optical regimes: (**a**–**c**) Southern Ocean; (**d**–**f**) South Pacific subtropical gyre; (**g**–**i**) Mediterranean Sea; (**j**–**l**) North Atlantic subpolar gyre—Irminger Sea.

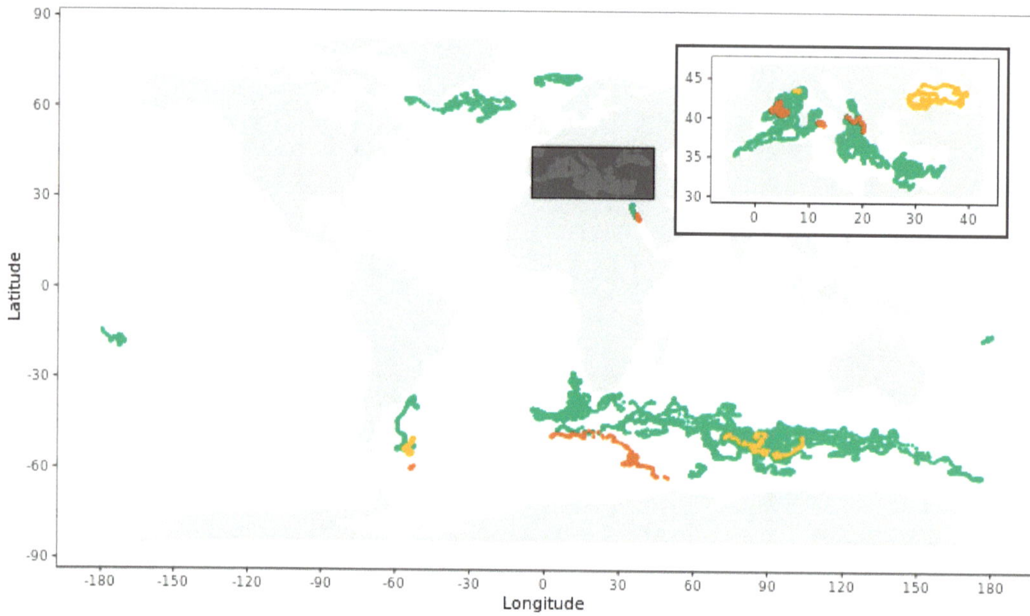


**Figure 8.** Radiometry profiles acquired by the 55 BGC-Argo floats with ancillary night profiles and drift measurements. Green dots: successfully corrected profiles with the DM-QC procedure; Orange dots: uncorrected profiles; Yellow dots: profiles corrected with alternative methods (see Supplementary Materials Section S2).

Regarding the remaining uncorrected 8 floats and 1043 radiometric profiles: 582 profiles from three floats (i.e., about 5% of the tested database) were corrected with alternative procedures specifically developed for the array of 76 floats with an insufficient number of night profiles or drift measurements (Supplementary Materials Section S2), while 461 profiles from five floats (i.e., about 4% of the tested database) were not corrected. Correction was made with alternative procedures when the ancillary data (most often night profiles) were not good enough to confidently apply the procedure described here, correction was abandoned when the alternative methods also failed.

Overall, the DM-QC procedure to correct the sensor dark signal systematically succeeded for all tested floats with at least four night profiles collected over the float lifetime (Figure 9).

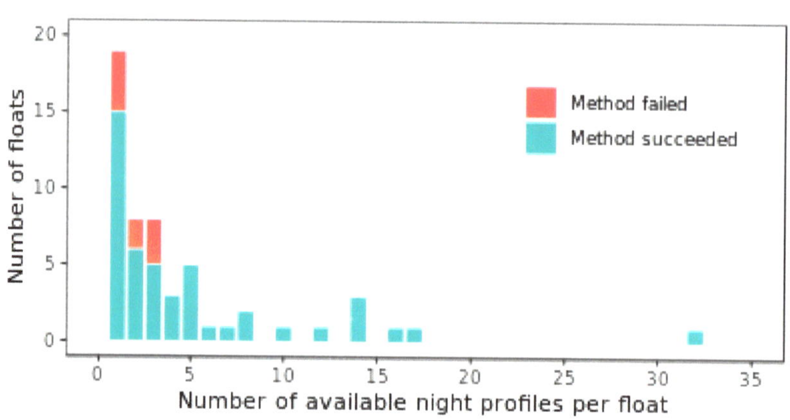


**Figure 9.** Number of floats with dark measurements successfully corrected for the four radiometric channels as a function of available night profiles.

Nevertheless, the majority of floats had three or fewer associated night profiles over their lifetime, and the correction we implemented was still successful in most of those cases. As the average lifespan of a float is expected to be four years [3], our results thus implies that each float equipped with radiometers must acquire one night profile per year, preferably during moonless nights and when the temperature range between the surface and 1000 dbar parking depth is the largest.

## 5. Discussion and Conclusions

To quality-control the large amount of radiometric profiles acquired by BGC-Argo floats, real-time [32] and near real-time quality-control procedures [28] have been proposed. While the method proposed by Poteau et al. [32] was mainly verifying the range of measured values, Organelli et al. [28] proposed protocols for the qualification of radiometric profiles to specifically use in ocean optics science and remote sensing applications e.g., for the derivation of the diffuse attenuation coefficient $K_d$ which is a key quantity for bio-optical and biogeochemical studies [4]. With this aim, their method was not focusing on the issues addressed here (i.e., sensor's dark dependence on temperature and aging) but rather on how the environment (presence of clouds, wave focusing at the surface) drives departures of the profile with respect to an expected monotonic decrease of irradiance with depth. Moreover, the scientific exploitation of the quality controlled radiometric profiles according to Organelli et al. [28] was restricted to the upper layer (i.e., first penetration depth [45]), mainly because some inconsistencies likely due to sensor dark's temperature-dependence issues were noticed in the deepest part.

The method proposed here offers a pragmatic way to identify and correct BGC-Argo radiometric profiles for sensor dark's aging and temperature-dependence issues, by acquiring one night profile per year and daily dark measurements at the 1000 dbar parking depth. These new protocols will allow to extend the range of exploitable measurements and, ultimately, enhance their use among the international biogeochemical community. Yet, we also recommend a technological upgrade of radiometers installed on floats with a probe to directly monitoring the internal temperature at which the sensor operates, which has only been modelled so far.

We must notice that the joint use of the DM-QC method here proposed with the one presented by Organelli et al. [28] represents an opportunity to generate a unique high-quality and interoperable radiometric dataset free of clouds and wave focusing/defocusing. Given the potential for the BGC-Argo network to expand [2,46], it can be expected that the resulting dataset, potentially increasing in near real-time, would allow addressing or readdressing key topics of applications in ocean optics the investigation of which was up to now suffering from limited data availability. The quality of the data could be further enhanced when also the impact of instrument tilt on measured values as well as the effect of bio-fouling that can occur [8] will be taken into account.

Among these ocean optics science topics, the understanding of regional and seasonal variability of $K_d$ with a higher degree of confidence must be refined along the water column [4]. Additionally, comparing such in-situ BGC-Argo float products with their satellite counterparts would allow the identification of the locations where bio-optical anomalies or nuances exist. This would represent a preliminary step to understand the causes of discrepancies and, as a consequence, possibly refine the retrieval algorithms for satellite products in some areas.

The possible derivation of radiometry with depth over the whole vertical dimension is expected to provide high resolution $K_d$ profiles that will be useful to address the link between surface remotely-sensed properties and their vertical variability according to region and season. Such data could in turn allow to re-evaluate and possibly improve methods developed to retrieve the vertical profile of chlorophyll-a from simultaneous measurement of chlorophyll fluorescence and radiometry from floats, methods that were initially developed on a very small float dataset [47].

The improved accuracy of radiometric measurements with depth will also enhance their use across the biogeochemical and ecosystem model community. An improved accuracy is expected to support studies that assimilate irradiance data to model phytoplankton photosynthesis [26], especially at the most elevated depths where the deep chlorophyll maxima are observed and supported by small quantities of light.

When considering the DM corrected profiles over the whole tested database, the method we presented showed high and similar applicability for the three channels of downwelling irradiance as well as for PAR, thus suggesting potential applicability to hyperspectral radiometers. With the advent of future hyperspectral satellite missions [48], there is an increasing interest in in-situ hyperspectral optics. Profiling floats equipped with hyperspectral radiometers represent an especially cost-effective approach to evaluate satellite performances during the post-launching so-called commissioning phases (few months). Such technology would indeed allow the acquisition of numerous calibration/validation high-quality matchups in a limited period of time, provided that a significant fleet of dedicated floats [49] would be deployed in diverse environments with specific bio-optical status and atmospheric specificities. Additionally, hyperspectral measurements could possibly become a component of the standard BGC-Argo fleet offering the possibility to refine the detection and quantification of optically significant substances (phytoplankton communities, detritus, mineral substance, colored dissolved organic matter).

Finally, it should be noticed that with the increasing development of robotic observation systems, a fleet of sensors can now be deployed and operated globally which definitely will change our way to look at data and qualify them. Working with a dense dataset acquired from multiple-a priori identical and interoperable-instruments will indeed allow us to identify sensor issues that would be difficult to discover on a case-by-case analysis [43]. In this respect the BGC-Argo network represents a unique platform to help in improving sensor performances for the benefit of other observation systems.

**Supplementary Materials:** The following are available online at https://www.mdpi.com/article/10.3390/s21186217/s1.

**Author Contributions:** Conceptualization, Q.J., E.O. and H.C.; methodology, Q.J., E.O., N.B., X.X., C.S., E.B., A.P., E.L. and M.C.; software, Q.J.; validation, Q.J., E.O. and H.C.; formal analysis, Q.J.; investigation, Q.J. and E.O. and H.C.; resources, C.S. and A.P.; data curation, Q.J., C.S. and A.P.; writing—original draft preparation, Q.J., E.O. and H.C.; writing—review and editing, all; visualization, Q.J.; supervision, E.O., C.S. and H.C.; project administration, F.D. and H.C.; funding acquisition, F.D. and H.C. All authors have read and agreed to the published version of the manuscript.

**Funding:** This research was funded by EUROPEAN RESEARCH COUNCIL, grant numbers 246777 (remOcean project) and 834177 (REFINE project); EUROPEAN UNION'S HORIZON 2020 research and innovation program, grant numbers 2014-633211 (AtlantOS project) and 824131 (EA-RISE project); AGENCE NATIONALE DE LA RECHERCHE in the framework of the French "Equipement d'avenir" program, grant number ANR J11R107-F (NAOS project); BNP PARIBAS FOUNDATION (SOCLIM project); CNES-TOSCA and LEFE-GMMC (BGC-Argo France program); SECOND INSTITUTE OF OCEANOGRAPHY, MNR, grant numbers QNYC1702 and 14283. The APC was funded by EUROPEAN RESEARCH COUNCIL through the REFINE project (grant number 834177).

**Institutional Review Board Statement:** Not applicable.

**Informed Consent Statement:** Not applicable.

**Data Availability Statement:** The data used in this study are openly available at https://doi.org/10.17882/42182 (accessed on 13 September 2021).

**Acknowledgments:** The technical staff at the French GDAC (Coriolis) is acknowledged for their invaluable support.

**Conflicts of Interest:** The authors declare no conflict of interest. The funders had no role in the design of the study; in the collection, analyses, or interpretation of data; in the writing of the manuscript, or in the decision to publish the results.

**Computer Code and Software:** The computed code to implement the protocols presented in this study is openly available at https://github.com/qjutard/radiometry_QC (version 1.03, accessed on 13 September 2021).

## Nomenclature

| Symbol | Definition |
|---|---|
| $E_d$ | Downwelling irradiance |
| PAR | Photosynthetically Available Radiation |
| $Im$ | Immersion coefficient |
| $a_0$; $a_1$ | Calibration coefficients |
| DC | Dark counts |
| $t$ | Time |
| $T_s$ | Sensor internal temperature |
| $k$ | Rate of change of the sensor temperature |
| $T_w$ | Water temperature |
| $\Delta t$ | Response delay of the sensor temperature to the water temperature |
| $c$ | Ascending speed of floats (assumed constant) |
| $T_s^*$ | Sensor temperature delayed by $\Delta t$ |
| $T_{s_n}$ | Discretized sensor temperature |
| $T_{w_n}$ | Water temperature measurements, sorted from the deepest to the shallowest |
| $T_{s_n}^*$ | Discretized delayed sensor temperature, follows the water temperature measurements axis |
| $t_n$ | Discretized time corresponding to water temperature measurements |
| $Pw_n$ | Pressure measurements associated to water temperature measurements |
| $Ps_n$ | Pressure axis associated to $T_{s_n}$ |
| $E_{d_{meas}}$ | Measured irradiance |
| $E_{d_{real}}$ | Real irradiance that would be obtained with a perfect sensor |
| $h$ | Slope error introduced by the temperature and aging effects |
| $\varepsilon$ | Sensor noise |
| $f(T_s)$ | Error offset caused by the sensor temperature being different from calibration |
| $g(t)$ | Error offset caused by sensor aging over time |
| $E_{d_{meas}}^*$ | Measured irradiance, fitted to $T_s$ and $t$ |
| $Ad$, $Bd$, $Cd$, $Qd$ | Coefficients in the fit of drift measurements to $T_s$ and $t$ |
| $E_{d_{5C}}$ | Measured irradiance in drift, projected on the $T_s = 5$ °C plane along the $E_{d_{meas}}^*$ fit |
| $E_{d_{5C}}^*$ | $E_{d_{meas}}^*$ projected on the $T_s = 5$ °C plane along the $E_{d_{meas}}^*$ fit |
| $E_{d_{night}}$ | Irradiance measurements in night profiles, corrected for sensor aging |
| $E_{d_{night}}^*$ | $E_{d_{night}}$, fitted to $T_s$ |
| $At$<br>$Bt$ | Coefficients in the fit of night measurements to $T_s$ |
| $E_{d_{corr}}$ | Irradiance corrected for the effects of temperature and aging on the dark signal |
| $A$, $B$, $C$, $Q$ | Coefficients in the full expression of the irradiance correction |
| $\sigma_{E_d}$ | Error associated to $E_d$ |
| $NEI_{E_d}$ | Noise Equivalent Irradiance |
| $ER_{E_d}$ | Relative Error |

## References

1. Johnson, K.S.; Claustre, H. Bringing Biogeochemistry into the Argo Age. *Eos* **2016**, 97. [CrossRef]
2. Claustre, H.; Johnson, K.S.; Takeshita, Y. Observing the Global Ocean with Biogeochemical-Argo. *Annu. Rev. Mar. Sci.* **2020**, 12, 23–48. [CrossRef] [PubMed]
3. Biogeochemical-Argo Planning Group. *The Scientific Rationale, Design and Implementation Plan for a Biogeochemical-Argo Float Array*; Ifremer: Brest, France, 2016.

4. Organelli, E.; Claustre, H.; Bricaud, A.; Barbieux, M.; Uitz, J.; D'Ortenzio, F.; Dall'Olmo, G. Bio-Optical Anomalies in the World's Oceans: An Investigation on the Diffuse Attenuation Coefficients for Downward Irradiance Derived from Biogeochemical Argo Float Measurements. *J. Geophys. Res. Oceans* **2017**, *122*, 3543–3564. [CrossRef]
5. Xing, X.; Claustre, H.; Wang, H.; Poteau, A.; D'Ortenzio, F. Seasonal Dynamics in Colored Dissolved Organic Matter in the Mediterranean Sea: Patterns and Drivers. *Deep-Sea Res. Part I Oceanogr. Res. Pap.* **2014**, *83*, 93–101. [CrossRef]
6. Organelli, E.; Claustre, H. Small Phytoplankton Shapes Colored Dissolved Organic Matter Dynamics in the North Atlantic Subtropical Gyre. *Geophys. Res. Lett.* **2019**, *46*, 12183–12191. [CrossRef] [PubMed]
7. Gerbi, G.P.; Boss, E.; Werdell, P.J.; Proctor, C.W.; Haëntjens, N.; Lewis, M.R.; Brown, K.; Sorrentino, D.; Zaneveld, J.R.V.; Barnard, A.H.; et al. Validation of Ocean Color Remote Sensing Reflectance Using Autonomous Floats. *J. Atmos. Ocean. Technol.* **2016**, *33*, 2331–2352. [CrossRef]
8. Organelli, E.; Barbieux, M.; Claustre, H.; Schmechtig, C.; Poteau, A.; Bricaud, A.; Boss, E.; Briggs, N.; Dall'Olmo, G.; D'Ortenzio, F.; et al. Two Databases Derived from BGC-Argo Float Measurements for Marine Biogeochemical and Bio-Optical Applications. *Earth Syst. Sci. Data* **2017**, *9*, 861–880. [CrossRef]
9. Wojtasiewicz, B.; Hardman-Mountford, N.J.; Antoine, D.; Dufois, F.; Slawinski, D.; Trull, T.W. Use of Bio-Optical Profiling Float Data in Validation of Ocean Colour Satellite Products in a Remote Ocean Region. *Remote Sens. Environ.* **2018**, *209*, 275–290. [CrossRef]
10. Leymarie, E.; Penkerc'h, C.; Vellucci, V.; Lerebourg, C.; Antoine, D.; Boss, E.; Lewis, M.R.; D'Ortenzio, F.; Claustre, H. ProVal: A New Autonomous Profiling Float for High Quality Radiometric Measurements. *Front. Mar. Sci.* **2018**, *5*, 18. [CrossRef]
11. Volpe, G.; Colella, S.; Brando, V.E.; Forneris, V.; La Padula, F.; Di Cicco, A.; Sammartino, M.; Bracaglia, M.; Artuso, F.; Santoleri, R. Mediterranean Ocean Colour Level 3 Operational Multi-Sensor Processing. *Ocean Sci.* **2019**, *15*, 127–146. [CrossRef]
12. Xing, X.; Boss, E.; Zhang, J.; Chai, F. Evaluation of Ocean Color Remote Sensing Algorithms for Diffuse Attenuation Coefficients and Optical Depths with Data Collected on BGC-Argo Floats. *Remote Sens.* **2020**, *22*, 2367. [CrossRef]
13. Xing, X.; Boss, E. Chlorophyll-Based Model to Estimate Underwater Photosynthetically Available Radiation for Modeling, In-Situ, and Remote-Sensing Applications. *Geophys. Res. Lett.* **2021**, *48*, e2020GL092189. [CrossRef]
14. Dall'Olmo, G.; Dingle, J.; Polimene, L.; Brewin, R.J.W.; Claustre, H. Substantial Energy Input to the Mesopelagic Ecosystem from the Seasonal Mixed-Layer Pump. *Nat. Geosci.* **2016**, *9*, 820–823. [CrossRef] [PubMed]
15. Lacour, L.; Briggs, N.; Claustre, H.; Ardyna, M.; Dall'Olmo, G. The Intraseasonal Dynamics of the Mixed Layer Pump in the Subpolar North Atlantic Ocean: A Biogeochemical-Argo Float Approach. *Glob. Biogeochem. Cycles* **2019**, *33*, 266–281. [CrossRef]
16. Kheireddine, M.; Dall'Olmo, G.; Ouhssain, M.; Krokos, G.; Claustre, H.; Schmechtig, C.; Poteau, A.; Zhan, P.; Hoteit, I.; Jones, B.H. Organic Carbon Export and Loss Rates in the Red Sea. *Glob. Biogeochem. Cycles* **2020**, *34*, e2020GB006650. [CrossRef]
17. Mignot, A.; Claustre, H.; Uitz, J.; Poteau, A.; D'Ortenzio, F.; Xing, X. Understanding the Seasonal Dynamics of Phytoplankton Biomass and the Deep Chlorophyll Maximum in Oligotrophic Environments: A Bio-Argo Float Investigation. *Glob. Biogeochem. Cycles* **2014**, *28*, 856–876. [CrossRef]
18. Lacour, L.; Ardyna, M.; Stec, K.F.; Claustre, H.; Prieur, L.; Poteau, A.; Ribera D'Alcala, M.; Iudicone, D. Unexpected Winter Phytoplankton Blooms in the North Atlantic Subpolar Gyre. *Nat. Geosci.* **2017**, *10*, 836–839. [CrossRef]
19. Mayot, N.; D'Ortenzio, F.; Taillandier, V.; Prieur, L.; Pasqueron de Fommervault, O.; Claustre, H.; Bosse, A.; Testor, P.; Conan, P. Physical and Biogeochemical Controls of the Phytoplankton Blooms in North Western Mediterranean Sea: A Multiplatform Approach Over a Complete Annual Cycle (2012–2013 DEWEX Experiment). *J. Geophys. Res. Oceans* **2017**, *122*, 9999–10019. [CrossRef]
20. Barbieux, M.; Uitz, J.; Bricaud, A.; Organelli, E.; Poteau, A.; Schmechtig, C.; Gentili, B.; Obolensky, G.; Leymarie, E.; Penkerc'h, C.; et al. Assessing the Variability in the Relationship between the Particulate Backscattering Coefficient and the Chlorophyll *a* Concentration From a Global Biogeochemical-Argo Database. *J. Geophys. Res. Oceans* **2018**, *123*, 1229–1250. [CrossRef]
21. Mignot, A.; Ferrari, R.; Claustre, H. Floats with Bio-Optical Sensors Reveal What Processes Trigger the North Atlantic Bloom. *Nat. Commun.* **2018**, *9*, 190. [CrossRef] [PubMed]
22. Barbieux, M.; Uitz, J.; Gentili, B.; Pasqueron de Fommervault, O.; Mignot, A.; Poteau, A.; Schmechtig, C.; Taillandier, V.; Leymarie, E.; Penkerc'h, C.; et al. Bio-Optical Characterization of Subsurface Chlorophyll Maxima in the Mediterranean Sea from a Biogeochemical-Argo FLoat Database. *Biogeosciences* **2019**, *16*, 1321–1342. [CrossRef]
23. Kubryakov, A.A.; Mikaelyan, A.S.; Stanichny, S.V. Summer and Winter Coccolithophore Blooms in the Black Sea and Their Impact on Production of Dissolved Organic Matter from Bio-Argo Data. *J. Mar. Syst.* **2019**, *199*, 103220. [CrossRef]
24. Randelhoff, A.; Lacour, L.; Marec, C.; Leymarie, E.; Lagunas, J.; Xing, X.; Darnis, G.; Penkerc'h, C.; Sampei, M.; Fortier, L.; et al. Arctic Mid-Winter Phytoplankton Growth Revealed by Autonomous Profilers. *Sci. Adv.* **2020**, *6*, eabc2678. [CrossRef]
25. Ricour, F.; Capet, A.; D'Ortenzio, F.; Delille, B.; Grégoire, M. Dynamics of the Deep Chlorophyll Maximum in the Black Sea as Depicted by BGC-Argo Floats. *Biogeosciences* **2021**, *18*, 755–774. [CrossRef]
26. Terzić, E.; Lazzari, P.; Organelli, E.; Solidoro, C.; Salon, S.; D'Ortenzio, F.; Conan, P. Merging Bio-Optical Data from Biogeochemical-Argo Floats and Models in Marine Biogeochemistry. *Biogeosciences* **2019**, *16*, 2527–2542. [CrossRef]
27. Lazzari, P.; Salon, S.; Terzić, E.; Gregg, W.W.; D'Ortenzio, F.; Vellucci, V.; Organelli, E.; Antoine, D. Assessment of the Spectral Downward Irradiance at the Surface of the Mediterranean Sea Using the Radiative Ocean-Atmosphere Spectral Irradiance Model (OASIM). *Ocean Sci.* **2021**, *17*, 675–697. [CrossRef]

28. Organelli, E.; Claustre, H.; Bricaud, A.; Schmechtig, C.; Poteau, A.; Xing, X.; Prieur, L.; D'Ortenzio, F.; Dall'Olmo, G.; Vellucci, V. A Novel Near-Real-Time Quality-Control Procedure for Radiometric Profiles Measured by Bio-Argo Floats: Protocols and Performances. *J. Atmos. Ocean. Technol.* **2016**, *33*, 937–951. [CrossRef]
29. IOCCG Protocol Series. *Protocols for Satellite Ocean Colour Data Validation: In Situ Optical Radiometry. Vol. 3.0*; International Ocean Colour Coordinating Group (IOCCG): Dartmouth, NS, Canada, 2019.
30. Xing, X.; Lagunas-Morales, J. Laboratory Results on the Dependence of Dark Current upon Environmental Temperature Variability for Satlantic's OCR504 Radiometers. In Proceedings of the Optical Precision Manufacturing, Testing, and Applications, Beijing, China, 22–24 May 2018.
31. Argo Data Management. *Argo User's Manual V3.4*; Ifremer: Brest, France, 2021.
32. Poteau, A.; Organelli, E.; Boss, E.; Xing, X. *Quality Control for BGC-Argo Radiometry*; Ifremer: Brest, France, 2019.
33. Mueller, J.L.; Fargion, G.S.; McClain, C.R.; Mueller, J.L.; Morel, A.; Frouin, R.; Davis, C.; Arnone, R.; Carder, K.; Steward, R.G.; et al. *Ocean Optics Protocols For Satellite Ocean Color Sensor Validation, Revision 4, Volume III: Radiometric Measurements and Data Analysis Protocols*; Goddard Space Flight Space Center: Greenbelt, MD, USA, 2003; p. 84.
34. Białek, A.; Vellucci, V.; Gentili, B.; Antoine, D.; Gorroño, J.; Fox, N.; Underwood, C. Monte Carlo-Based Quantification of Uncertainties in Determining Ocean Remote Sensing Reflectance from Underwater Fixed-Depth Radiometry Measurements. *J. Atmos. Ocean. Technol.* **2020**, *37*, 177–196. [CrossRef]
35. Bittig, H.C.; Maurer, T.L.; Plant, J.N.; Schmechtig, C.; Wong, A.P.S.; Claustre, H.; Trull, T.W.; Udaya Bhaskar, T.V.S.; Boss, E.; Dall'Olmo, G.; et al. A BGC-Argo Guide: Planning, Deployment, Data Handling and Usage. *Front. Mar. Sci.* **2019**, *6*, 502. [CrossRef]
36. Wong, A.; Keeley, R.; Carval, T.; A.D.M. Team. *Argo Quality Control Manual for CTD and Trajectory Data*; Ifremer: Brest, France, 2021.
37. SATLANTIC. *Operation Manual for the OCR-504*; SATLANTIC Operation Manual; Satlantic LP: Halifax, NS, Canada, 2013; p. 66.
38. Shockley, W. The Theory of p-n Junctions in Semiconductors and p-n Junction Transistors. *Bell Syst. Tech. J.* **1949**, *28*, 435–489. [CrossRef]
39. Kuusk, J. Dark Signal Temperature Dependence Correction Method for Miniature Spectrometer Modules. *J. Sens.* **2011**, *2011*. [CrossRef]
40. Möttus, M.; Sulev, M.; Baret, F.; Lopez-Lozano, R.; Reinart, A. Photosynthetically Active Radiation: Measurement and Modeling. In *SpringerReference*; Springer: Berlin/Heidelberg, Germany, 2011.
41. Hooker, S.B.; McLean, S.; Sherman, J.; Small, M.; Lazin, G.; Zibordi, G.; Brown, J.W. *Volume 17, The Seventh SeaWiFS Intercalibration Round-Robin Experiment (SIRREX-7), March 1999. SeaWiFS Postlaunch Technical Report Series*; Goddard Space Flight Space Center: Greenbelt, MD, USA, 2002; p. 74.
42. Voss, K.J.; McLean, S.; Lewis, M.; Johnson, C.; Flora, S.; Feinholz, M.; Yarbrough, M.; Trees, C.; Twardowski, M.; Clark, D. An Example Crossover Experiment for Testing New Vicarious Calibration Techniques for Satellite Ocean Color Radiometry. *J. Atmos. Ocean. Technol.* **2010**, *27*, 1747–1759. [CrossRef]
43. Poteau, A.; Boss, E.; Claustre, H. Particulate Concentration and Seasonal Dynamics in the Mesopelagic Ocean Based on the Backscattering Coefficient Measured with Biogeochemical-Argo Floats. *Geophys. Res. Lett.* **2017**, *44*, 6933–6939. [CrossRef]
44. Bellacicco, M.; Cornec, M.; Organelli, E.; Brewin, R.J.W.; Neukermans, G.; Volpe, G.; Barbieux, M.; Poteau, A.; Schmechtig, C.; D'Ortenzio, F.; et al. Global Variability of Optical Backscattering by Non-Algal Particles From a Biogeochemical-Argo Data Set. *Geophys. Res. Lett.* **2019**, *46*, 9767–9776. [CrossRef]
45. Gordon, H.R.; McCluney, W.R. Estimation of the Depth of Sunlight Penetration in the Sea for Remote Sensing. *Appl. Opt.* **1975**, *14*, 413. [CrossRef]
46. Roemmich, D.; Alford, M.H.; Claustre, H.; Johnson, K.; King, B.; Moum, J.; Oke, P.; Owens, W.B.; Pouliquen, S.; Purkey, S.; et al. On the Future of Argo: A Global, Full-Depth, Multi-Disciplinary Array. *Front. Mar. Sci.* **2019**, *6*, 439. [CrossRef]
47. Xing, X.; Morel, A.; Claustre, H.; Antoine, D.; D'Ortenzio, F.; Poteau, A.; Mignot, A. Combined Processing and Mutual Interpretation of Radiometry and Fluorimetry from Autonomous Profiling Bio-Argo Floats: Chlorophyll a Retrieval. *J. Geophys. Res.* **2011**, *116*, C06020. [CrossRef]
48. Werdell, P.J.; Behrenfeld, M.J.; Bontempi, P.S.; Boss, E.; Cairns, B.; Davis, G.T.; Franz, B.A.; Gliese, U.B.; Gorman, E.T.; Hasekamp, O.; et al. The Plankton, Aerosol, Cloud, Ocean Ecosystem Mission: Status, Science, Advances. *Bull. Am. Meteorol. Soc.* **2019**, *100*, 1775–1794. [CrossRef]
49. Jemai, A.; Wollschläger, J.; Voß, D.; Zielinski, O. Radiometry on Argo Floats: From the Multispectral State-of-the-Art on the Step to Hyperspectral Technology. *Front. Mar. Sci.* **2021**, *8*, 676537. [CrossRef]

*sensors*

MDPI

*Article*

# A Mosaic Method for Side-Scan Sonar Strip Images Based on Curvelet Transform and Resolution Constraints

**Ning Zhang, Shaohua Jin *, Gang Bian, Yang Cui and Liang Chi**

Department of Military Oceanography and Hydrography and Cartography, Dalian Naval Academy, Dalian 116018, China; nick_zn180@163.com (N.Z.); trighosts@163.com (G.B.); 13998435151@163.com (Y.C.); liangchi202107@163.com (L.C.)

* Correspondence: jsh_1978@163.com; Tel.: +86-17640356647

**Abstract:** Due to the complex marine environment, side-scan sonar signals are unstable, resulting in random non-rigid distortion in side-scan sonar strip images. To reduce the influence of resolution difference of common areas on strip image mosaicking, we proposed a mosaic method for side-scan sonar strip images based on curvelet transform and resolution constraints. First, image registration was carried out to eliminate dislocation and distortion of the strip images. Then, the resolution vector of the common area in two strip images were calculated, and a resolution model was created. Curvelet transform was then performed for the images, the resolution fusion rules were used for Coarse layer coefficients, and the maximum coefficient integration was applied to the Detail layer and Fine layer to calculate the fusion coefficients. Last, inverse Curvelet transform was carried out on the fusion coefficients to obtain images in the fusion area. The fusion images in multiple areas were then combined in the registered images to obtain the final image. The experiment results showed that the proposed method had better mosaicking performance than some conventional fusion algorithms.

**Keywords:** strip images mosaic; image resolution; curvelet transform; image fusion

## 1. Introduction

As the depth of global ocean exploration continues to increase, understanding the seafloor surface and near-surface is of great significance in the "digital ocean" and "transparent ocean" era. Currently, side-scan sonar is an important means to explore seafloor geomorphology [1], and side-scan sonar images provide important data for seafloor object identification, classification of seafloor sediments, and exploration of marine resources [2,3]. In order to obtain side-scan sonar images in the entire testing zone, the most important task is to mosaic the strip images, in addition to seafloor tracking, slant-range correction, gain correction, and geocoding [4]. Side-scan sonar is generally operated using towing cables, which leads to inaccurate location information. If the coordinate information is directly used to mosaic the images, there will be distortion in the images [5–8]. Currently, a large number of studies have been carried out to achieve a mosaic of object-level strip images that produce images with complete information and of high quality.

By dividing sonar strip images into paired objects and shadows, Daniel et al. [9] realized rigid registration of side-scan sonar images using a decision tree. Through region segmentation, Thisen et al. [10] extracted shadow areas from side-scan sonar images and calculated the displacement between two images using the cross-correlative method, thereby achieving rigid registration. Vandrish et al. [11] showed that the scale invariant feature transform (SIFT) algorithm can be used for registration of sonar images, although the accuracy was not ideal. Using correlation coefficients and mutual information as similarity parameters, Chailloux et al. [12] extracted a series of significantly correlated areas on adjacent strip images and calculated the global rigid transformation parameters and local elastic transformation parameters, thereby eventually realizing mosaic of adjacent strip images. Wang et al. [13] improved the pre-processing method of side-scan sonar images to extract feature points more accurately and effectively after preprocessing; they also proposed a sped up robust feature (SURF)-based elastic mosaic algorithm to achieve feature-level conformal mosaic of the images. Moreover, Cao et al. [14] used wavelet transform in a strip image mosaic, yet it required the 3D posture information of the side-scan sonar. Zhao et al. [15] extracted SURF features of the pre-processed strip images and then performed block registration, which achieved good mosaic results. To obtain sonar images of large-area seafloor surface, Zhao et al. [16] also proposed a side-scan image mosaicking method based on the coupling feature points of position constraints. In addition, He et al. [17] used the unsharp masking (USM) algorithm to enhance the side-scan images and the SURF algorithm for image mosaicking; experiments showed that their method effectively enhanced image features and increased the amount of image information, but the average gray values of the images were affected.

The above image mosaic algorithms primarily focused on the extraction and registration of features points of adjacent strip images, and most adopted the wavelet fusion algorithm after image registration, without further exploration for alternative image fusion algorithms. Due to the complex marine environment during ocean exploration, it is nearly impossible to ensure that the sonar images on one survey line are always better than those of an adjacent strip image. Therefore, it is necessary to take into account the differences in image resolution during strip image mosaicking and retain clear image information while screening necessary information in blurred images. To address this problem, we performed image fusion using curvelet transform, which can reveal more detailed information of strip images than wavelet transform. Then, the resolution of strip images was evaluated using a resolution weight model to constrain the curvelet transform, thereby achieving mosaicked strip images with better quality. The contents of this paper were arranged as follows: Section 2 mainly introduces seven different methods of resolution assessment, which would all be used in the calculation of resolution weight model; Section 3 mainly introduces the specific process of strip Mosaic method proposed in this paper; Section 4 uses the measured data to verify the feasibility of this method; and Section 5 contains the summary and prospects.

## 2. Image Resolution Assessment Methods

As an important data source of seafloor geomorphology, the resolution of side-scan sonar images directly determines the accuracy of target identification and seafloor sediments classification. The assessment of image quality can be divided into two types: subjective assessment and objective assessment [18,19]. Subjective assessment is mainly performed by trained professionals, whereas objective assessment uses mathematical models to measure the image resolution based on different indices. Thus, it is imperative to develop an objective assessment method that is in consistency with subjective assessment. Currently, common objective assessment methods can be divided into three categories according to the degree of use of reference images, i.e., full-reference quality assessment, reduced-reference quality assessment, and no-reference quality assessment [20]. Since there is no original reference image for side-scan sonar images, the no-reference quality assessment method was adopted in this study.

Image resolution is one of the most important image quality evaluation indexes and is the most important image parameter of sonar image. Therefore, the resolution of image became the main research object. A total of seven resolution assessment methods from four aspects will be introduced in this section. As the more classical parameter indexes in the assessment method, they measure the sharpness of the image from different aspects. Additionally, they will all be used in the calculation of resolution vector in Section 3, making the evaluation result more accurate and perfect.

### 2.1. Assessment Method Based on Image Gradient

Image gradient reflects the marginal information of images. The greater the gradient value is, the sharper the image edge and the clearer the image will be. Common gradient functions for evaluating image resolution include the following three types [21].

#### 2.1.1. Energy Gradient Function

The energy gradient of an image is the quadratic sum of the difference in grayscale value of adjacent pixels in the horizontal and vertical direction. The summation of energy gradient values of all pixels in the image is then taken as the function value. The function is shown in Equation (1):

$$F_{EG} = \sum_x \sum_y \left\{ [f(x+1,y) - f(x,y)]^2 + [f(x,y+1) - f(x,y)]^2 \right\} \tag{1}$$

where $x$ and $y$ are pixel coordinates, and $f(x,y)$ is the grayscale value of the pixel.

#### 2.1.2. Brenner Gradient Function

Brenner gradient function is relatively the easiest gradient assessment function [22]. It calculates the quadratic sum of the grayscale difference of two adjacent pixels, meaning a small calculation amount. Yet, it is sensitive to noise. The function is shown in Equation (2):

$$F_{Brenner} = \sum_x \sum_y [f(x+2,y) - f(x,y)]^2 \tag{2}$$

#### 2.1.3. Tenengrad Gradient Function

Krotkv et al. [23] used the Tenengrad gradient function as one of the assessment indexes of image resolution, the results of which were close to objective assessment results. In this method, the Sobel operator was first used to extract the horizontal and vertical gradient values of pixels, then the quadratic sum was compared with a threshold T. The gradient values of pixels greater than T were added to obtain the Tenengrad gradient function value. The function is shown in Equation (3):

$$F_{Tenengrad} = \sum_x \sum_y \left[ G(x,y)^2 \right] \tag{3}$$

where $G(x,y)$ is the gradient calculated by the Sobel operator, as shown in Equation (4):

$$G(x,y) = \sqrt{G_x^2(x,y) + G_y^2(x,y)} \tag{4}$$

where $G_x(x,y)$ and $G_y(x,y)$ represent the horizontal and vertical gradient values, respectively.

$$\begin{aligned} G_x(x,y) &= f(x,y) \otimes g_x \\ G_y(x,y) &= f(x,y) \otimes g_y \end{aligned} \tag{5}$$

where $\otimes$ is the convolution operator, and $g_x$ and $g_y$ represent the horizontal and vertical templates of the Sobel operator, respectively:

$$g_x = \begin{bmatrix} -1 & 0 & 1 \\ -2 & 0 & 2 \\ -1 & 0 & 1 \end{bmatrix} \quad g_y = \begin{bmatrix} -1 & -2 & -1 \\ 0 & 0 & 0 \\ 1 & 2 & 1 \end{bmatrix} \tag{6}$$

### 2.2. Assessment Method Based on Image Transform Domain

It is generally believed that a clear image contains more high-frequency components than a blurry image. Thus, some studies have attempted to transform the image to the frequency domain to perform image quality assessment [24].

#### 2.2.1. Discrete Fourier Transform (DFT)

As the most basic time–frequency transformation methods, DFT is widely used in resolution assessment. Specifically, 2D DFT is first performed on the image, and then the zero-frequency component is shifted to the matrix center, such that the frequency diffuses from the center to the periphery and from low frequency to high frequency. The spectrum values of corresponding pixels are weighted based on the distance to the central pixel, and the resolution assessment value is the weighted average of the spectrum values of all pixels [25,26]. The function of DFT-based image resolution assessment is shown in Equation (7) [27]:

$$F_{DFT} = \frac{1}{M \times N} \sum_{\mu=0}^{M-1} \sum_{\nu=0}^{N-1} \sqrt{\mu^2 + \nu^2} P(\mu, \nu) \tag{7}$$

where $M$ and $N$ are the image dimensions, $\sqrt{\mu^2 + \nu^2}$ represents the distance of a pixel to the central pixel, and $P(\mu, \nu)$ is the spectrum value of a pixel after DFT.

#### 2.2.2. Discrete Cosine Transform (DCT)

DFT-based resolution assessment methods have high sensitivity; however, they are computationally more demanding than DCT-based methods. In comparison, DCT has a general orthogonal transform property, and the base vector of DCT matrix could describe image features very well [28,29]. Therefore, by replacing DFT with DCT, the transform coefficient is changed into a real number, which reduces the computation while still obtaining the distribution of image frequency. The resolution assessment function based on DCT is shown in Equation (8):

$$F_{DCT} = \frac{1}{M \times N} \sum_{\mu=0}^{M-1} \sum_{\nu=0}^{N-1} (\lambda + \varphi) |C(\lambda, \varphi)| \tag{8}$$

where $C(\lambda, \varphi)$ is the spectrum value of a pixel after DCT.

### 2.3. Assessment Method Based on Entropy Function

The entropy of an image is an important index to measure the richness of image information. Shannon believed that the greater the entropy value, the richer information the image contains. During image resolution assessment, the clearer the image is, the more abundant grayscale distribution it has, and thus, the greater the entropy value is [30]. The definition of entropy function is shown in Equation (9):

$$F_{entropy} = \sum_{i=0}^{255} -p(i) \log_2 p(i) \tag{9}$$

where $p(i)$ is the probability of occurrence of every grayscale value.

### 2.4. Assessment Method Based on Variance Function

The variance function can represent the dispersion degree of the image grayscales. The smaller the range of grayscale, the smaller the variance is and the blurrier the image is, and vice versa [31]. The definition of variance function is shown in Equation (10):

$$F_{Var} = \sum_{x} \sum_{y} \{[f(x, y) - \varepsilon]^2\} \tag{10}$$

where $\varepsilon$ is the average grayscale value of the image, the definition of which is in Equation (11):

$$\varepsilon = \frac{1}{M \times N} \sum_{x} \sum_{y} f(x, y) \tag{11}$$

## 3. Strip Mosaic Method Based on Curvelet Transform and Resolution Constraints

### *3.1. Image Fusion Algorithm Based on Curvelet Transform*

To obtain a clear and continuous image that can reflect complete information of the entire testing zone, image fusion in the overlapping area of side-scan sonar strip images is required. Currently, there are three common image fusion methods, namely weighted average method, image pyramid method, and wavelet fusion method [32]. The wavelet fusion method is the most common side-scan sonar strip image mosaicking method. However, due to the limitations in algorithms, the wavelet transform can only obtain edge features in the horizontal and vertical directions, and the wavelet basis does not have the anisotropy property. Hence, it is unable to get close to the image texture features. To overcome the limitations in the wavelet transform and improve the quality of strip image mosaicking, the Curvelet transform was introduced in the current study.

The Curvetlet transform was first proposed by Candes and Donoho in 1999 [33] based on the Ridgelet transform. As a multi-resolution, band-pass, and directional multi-scale image analysis method, Curvelet transform has the three characteristics of an optimal image representation method proposed by the National Institute for Physiological Science, Japan [34]. Similar to wavelet transform, Curvelet transform calculates the correlation of spatial images using a group of base functions, thereby characterizing edges and curves at different angles. The main steps of image fusion based on Curvelet transform are as follows: Curvelet coefficients are first obtained from Curvelet decomposition of the image, the coefficients are then processed based on specific fusion rules, and lastly, inverse Curvelet transform is carried out on the fused coefficient to obtain the final fusion image [35,36].

The Curvelet coefficients are obtained using the equation below:

$$C(j, \theta, k_1, k_2) = \sum_{0 \le x \le M, 0 \le y \le N} f(x, y) \cdot \varphi_{j,\theta,k_1,k_2}(x, y) \tag{12}$$

where $f(x, y)$ is the input image, $M \times N$ are the image dimensions, $j$ is the scale, $\theta$ is the direction, $k_1, k_2$ is the spatial location of Curvelet, and $\varphi(x, y)$ represents the Curvelet function, which includes a group of base functions described by parameters $(j, \theta, k_1, k_2)$.

Different from the wavelet coefficients, the Curvelet coefficients include the low-frequency coefficient in the innermost layer (i.e., the Coarse layer), the mid-to-high frequency coefficient in the Detail layer, and the high-frequency coefficient in the outermost Fine layer. As the number of layers increases, the scale of the corresponding base function turns smaller, and there are more directions. Figure 1 shows a frequency-domain base division method. Each square in Figure 1 represents a scale, and there are five scales. The bigger the square, the higher the frequency, and the smaller the scale is; hence, more detailed information will be reflected. The radial lines represent the angles. At each scale, the angle division is different, and the higher the frequency is, the smaller the angle is.

From Jia et al. [37], the energy of coefficients is mainly concentrated in the low-frequency coefficient, and the energy gradually declines as the frequency increases. In other words, the low-frequency coefficient reflects the general trend of the image, whereas high-frequency coefficient reflects the outline and texture details of an image. By fusing the coefficients at various layers using different fusion rules, the fusion image coefficient can be obtained, and by performing inverse Curvelet transform of the fusion image coefficient, the fusion image is obtained.

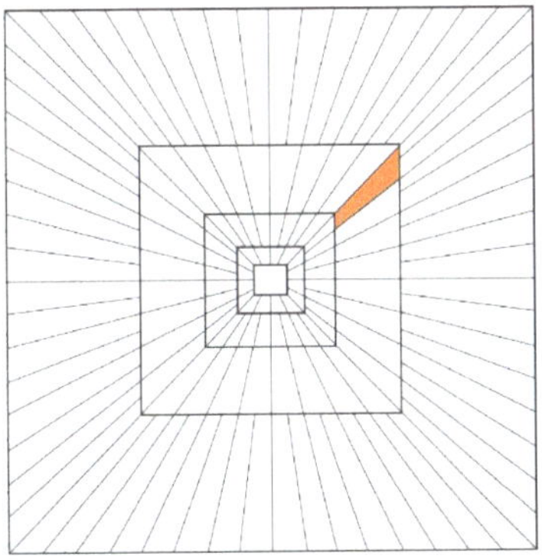

**Figure 1.** Frequency-Domain Base of Curvelet Transform.

### 3.2. *Strip Image Mosaicking Based on Curvelet Transform and Resolution Constraints*

Due to uncertainties in the marine environment during exploration, common areas in adjacent strip images might have large differences during actual measurement. Both strip images might have good quality, or one or both of them may not be good at all. The traditional side-scan strip image mosaicking algorithms do not take the image resolution into account. In order to ensure good mosaic results, a Curvelet coefficient fusion criterion based on the resolution weight model was proposed in the present study.

In Section 2, we have introduced seven different image resolution assessment methods, including energy gradient function, Brenner gradient function, Tenengrad gradient function, DFT, DCT, entropy function, and variance function. According to Li et al. [38] and Xie et al. [39], different resolution assessment methods may have different results for the same group of images. In other words, a single method is not able to assess the resolution of an image accurately. Hence, these seven resolution assessment methods were integrated in this study to build a resolution vector, and the image resolution was obtained based on probability and given weights.

The resolution vector $Q$, created based on the resolution value of the above seven methods, is shown in Equation (13):

$$Q = [F_{EG}, F_{Brenner}, F_{Tenengrad}, F_{DFT}, F_{DCT}, F_{entropy}, F_{Var}] \tag{13}$$

Since the resolution index in each method has a positive relationship with the image resolution, the resolution weight is obtained by comparing the resolution vectors of image 1 and image 2, $Q_1, Q_2$, respectively.

$$Ratio = \frac{sum(Q_1 \geq Q_2)}{7} \tag{14}$$

where the resolution weight *Ratio* represents the probability of an image having better resolution than the other image. Thus, it was taken as the fusion rule in the Coarse layer of Curvelet transform, as shown in Equation (15).

$$C_{Coarse\_fusion} = Ratio \cdot C_{Coarse\_1} + (1 - Ratio) \cdot C_{Coarse\_2} \tag{15}$$

where $C_{\text{Coarse\_}fusion}$, $C_{\text{Coarse\_1}}$ and $C_{\text{Coarse\_2}}$ represent the coefficient in the Coarse layer after fusion and that of image 1 and image 2, respectively.

In order to fully show the texture and details of the image, the maximum coefficient fusion approach was adopted to process the Detail layer and Fine layer coefficients, as shown in Equation (16):

$$\begin{aligned} C_{Detail\_fusion}(x,y) &= Max\{|C_{Detail\_1}(x,y)|, |C_{Detail\_2}(x,y)|\} \\ C_{Fine\_fusion}(x,y) &= Max\{|C_{Fine\_1}(x,y)|, |C_{Fine\_2}(x,y)|\} \end{aligned} \quad (16)$$

where $C_{Detail\_fusion}$, $C_{Detail\_1}$ and $C_{Detail\_2}$ represent the coefficient in the Detail layer after fusion and that of image 1 and image 2, respectively. $C_{Fine\_fusion}$, $C_{Fine\_1}$ and $C_{Fine\_2}$ represent the coefficient of the Fine layer after fusion and that of image 1 and image 2, respectively. Figure 2 shows the flowchart of the proposed mosaic method based on Curvelet transform and resolution constraints.

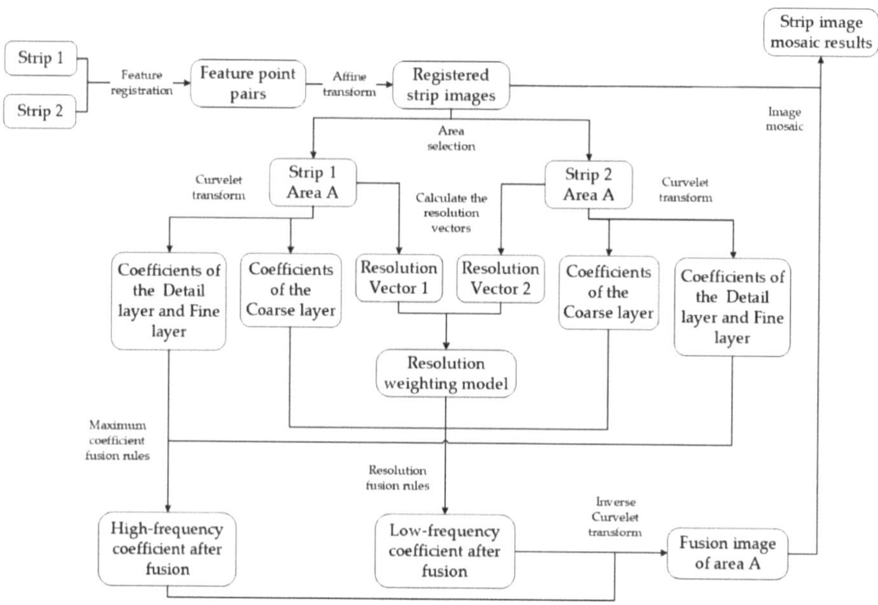


**Figure 2.** Flowchart of the proposed method.

1. Extract and match feature points of adjacent strip images and obtain registered mosaic strips using the affine transformation.
2. Select the common area A from two strip images.
3. Perform Curvelet transform for two images to obtain the coefficients in the Coarse layer, Detail layer, and Fine layer.
4. Calculate the resolution vectors of the two images to obtain the corresponding resolution weight.
5. Fuse the Coarse layer coefficients using resolution fusion rules to obtain the low-frequency coefficients. Fuse the Detail layer and Fine layer coefficients using the maximum coefficient fusion rules to obtain the high-frequency coefficients.
6. Perform inverse Curvelet transform on the fusion coefficients to obtain the fusion image in area A, which is then mosaicked to the registered strip images.
7. Repeat steps 2–6 until the whole mosaic image is obtained.

In traditional mosaic algorithms for strip images, there are various problems, such as inconsistent resolution of adjacent strip images and image distortion. In this study, we proposed a mosaic method for strip images based on Curvelet transform and resolution constraints, which produced mosaic images with complete information and high quality.

## 4. Experiment and Results

To verify the effectiveness of the proposed image mosaicking method, image data collected in 2019 using the Klein4000 side-scan sonar in Jiaozhou Bay, Qingdao, Shandong Province, China was used in the experiment. The water depth of the survey area is approximately 30–40 m. The overlapping rate of adjacent strip images is 50%. After preprocessing, such as seafloor tracking, slant-range correction, gray level equalization, noise suppression, gain correction, and geocoding, a group of strip image pairs with obvious common features were selected, as shown in Figure 3.

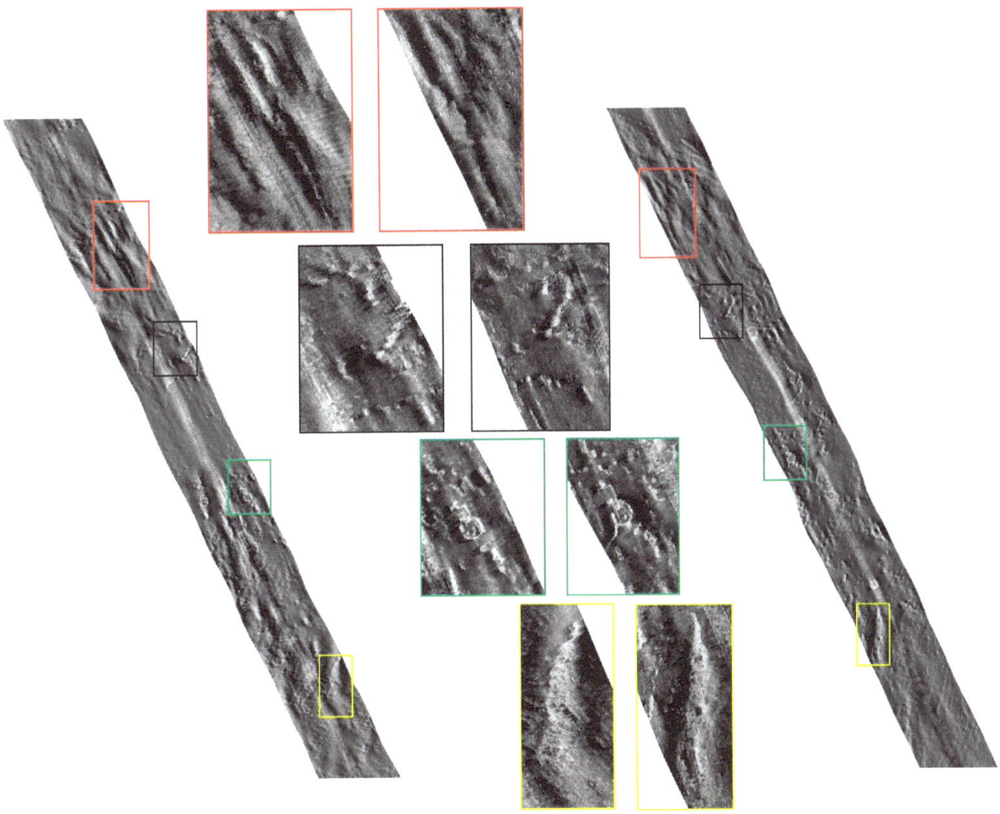

**Figure 3.** Two strips used for verification. Four image pairs with obvious common features were selected.

Figure 4a shows a mosaic image calculated based on geographic coordinate information. As can be seen, there is obvious dislocation and distortion. According to the steps of our method, the feature points in the strip images were extracted and matched, as shown in Figure 4b. Figure 4c shows a registered strip image after affine transformation. Based on the results, the distortion and dislocation were eliminated after image registration, resulting in good visual effects and laying a solid foundation for image fusion in the next step.

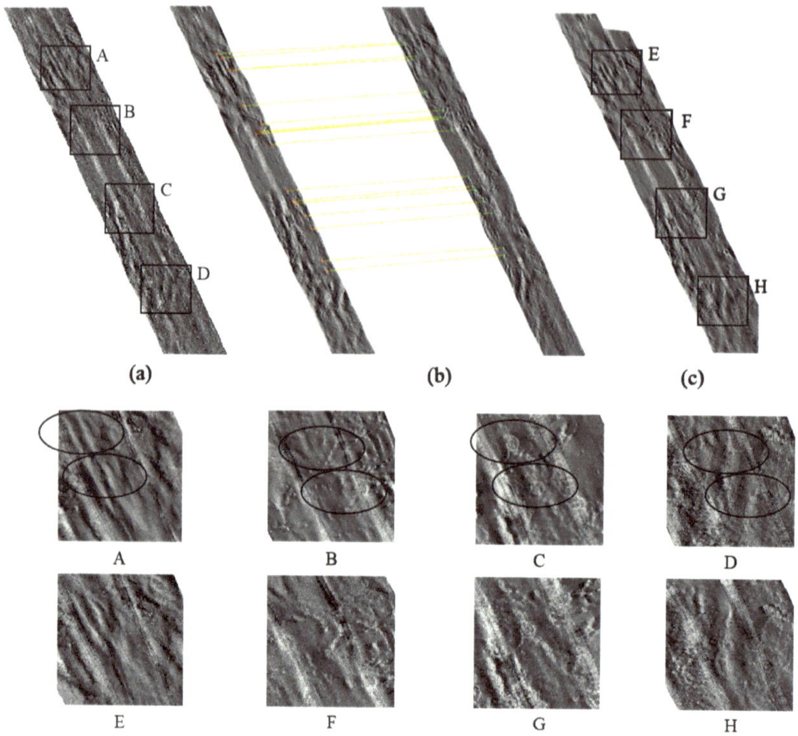


**Figure 4.** Strip image registration. A–D and E–H are four areas selected from (**a**) and (**c**) respectively. (**b**) shows the registration process of strips. It can be seen that there was significant dislocation in A–D. After strip registration, the dislocation effect largely disappeared in E–H.

To effectively select the fusion area and ensure the integrity of the selected features, the whole survey area was first rotated counterclockwise for a certain angle, such that the survey line was approximately along the vertical direction [40]. Another reason to rotate the strips is that a series of subsequent steps, such as Curvelet transform and image fusion, require regular rectangles. After image mosaicking, it was rotated back to the original direction. Areas 1–3 were selected, and the sonar images of two strips in these areas are shown in Figure 5.

Taking Area 1 as an example, the proposed algorithm was used to process two strips in the area. First, the coefficients in the Coarse, Detail, and Fine layers were extracted using Curvelet transform. The coefficient structure is shown in Table 1.

In both strips, Area 1 has the same dimensions of $923 \times 166$. Five layer decomposition was carried out. As shown in Table 1, the dimension of the coefficient matrix increases with the increase in scale. The larger the scale in spatial domain, the smaller the scale in frequency domain, and the more detailed the description of high frequency information.

Then, the resolution vectors of two strips in Area 1 were calculated, and the results are shown in Table 2.

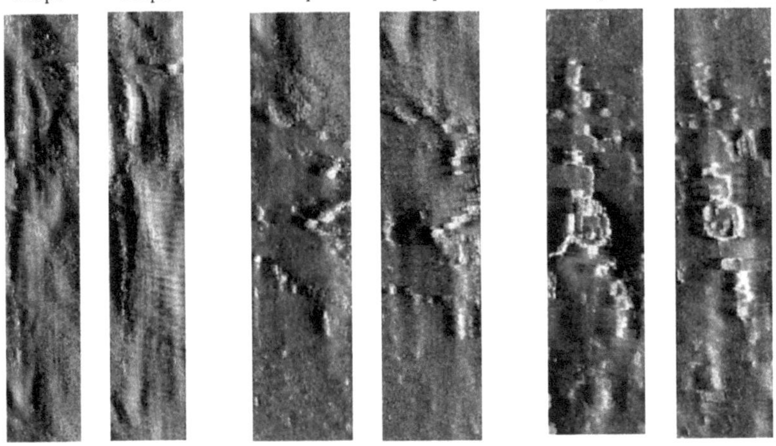


**Figure 5.** Selected fusion areas.

**Table 1.** Structure of Curvelet transform coefficients.

| Layer | Scale Coefficient | Number of Directions | Matrix Dimensions |
|---|---|---|---|
| Coarse | C{1} | 1 | 77 × 13 |
| Detail | C{2} | 16 | 62 × 14 57 × 14<br>77 × 11 77 × 10 |
| | C{3} | 32 | 120 × 14 115 × 15<br>115 × 14 115 × 15<br>77 × 22 78 × 21<br>77 × 21 |
| | C{4} | 32 | 241 × 29 231 × 28<br>231 × 29 154 × 44<br>155 × 42 155 × 42<br>154 × 42 |
| Fine | C{5} | 1 | 923 × 166 |

**Table 2.** Resolution vectors of two strips in Area 1.

| | $F_{EG}$ | $F_{Brenner}$ | $F_{Tenengrad}$ | $F_{DFT}$ | $F_{DCT}$ | $F_{entropy}$ | $F_{Var}$ |
|---|---|---|---|---|---|---|---|
| $Q_1$ | $1.6 \times 10^9$ | $7.2 \times 10^8$ | $6.0 \times 10^8$ | $2.5 \times 10^8$ | $4.3 \times 10^3$ | $2.6 \times 10^6$ | 7.303 |
| $Q_2$ | $2.2 \times 10^9$ | $1.4 \times 10^8$ | $1.0 \times 10^8$ | $3.4 \times 10^8$ | $6.2 \times 10^3$ | $3.3 \times 10^6$ | 6.886 |

$Q_1$, $Q_2$ denote the resolution vectors of Strip 1 and Strip 2, respectively. Additionally, the resolution weight ratio, computed according to Equation (14), is 0.1428.

Then, using the proposed algorithm, the coefficients in the Coarse layer of the two images were fused based on the resolution fusion rule, and the coefficients in the Detail and Fine layers of the two images were fused using the maximum coefficient fusion approach, thereby obtaining the low-frequency and high-frequency coefficients of the fused image. Lastly, the fused image of Area 1 was obtained via inverse Curvelet transform.

In order to verify the rationality of the resolution fusion rule proposed in this paper, the resolution fusion rule, the mean fusion rule, and the maximum fusion rule are used to combine the five layer coefficients of the two images obtained by the Curvelet decomposition, respectively. As shown in Figure 6, 19 combinations of fusion coefficients were obtained.

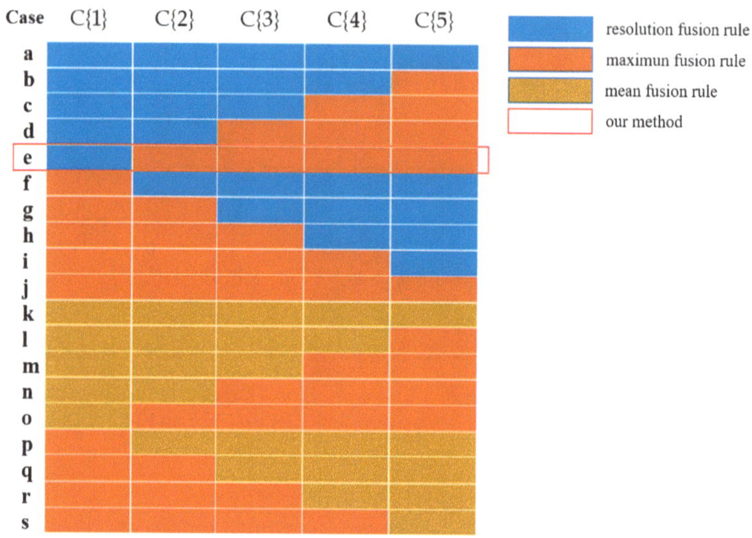


**Figure 6.** Combination diagram of fusion rules about Curvelet transform coefficients.

Then, the fusion coefficients of each group were inversely transformed to obtain fusion images.

The information entropy, average gradient, and spatial frequency were used as evaluation indices of the fusion results. The information entropy reveals the amount of information contained in the image, and the greater the entropy, the better the fusion result; the average gradient reflects the image's contrast expression of small details, and the greater the average gradient, the higher the image fusion quality; the spatial frequency represents the overall activity of the image in spatial domain, and the higher the spatial frequency, the better the fusion result. Table 3 shows the three indices of each combination, and Figure 7 shows the line chart of the analysis results.

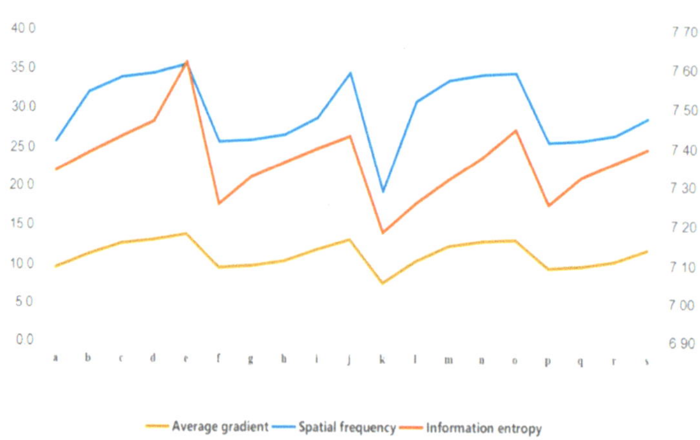


**Figure 7.** Line chart of the fusion coefficients analysis results.

Table 3. Comparison of fusion effects in different combinations.

| Case | Information Entropy | Average Gradient | Spatial Frequency |
|---|---|---|---|
| a | 7.3376 | 9.4983 | 25.5944 |
| b | 7.3826 | 11.1938 | 31.9010 |
| c | 7.4250 | 12.5927 | 33.7882 |
| d | 7.4637 | 13.0406 | 34.3080 |
| e [1] | 7.6156 | 13.7586 | 35.4629 |
| f | 7.2523 | 9.4976 | 25.5687 |
| g | 7.3222 | 9.7393 | 25.8062 |
| h | 7.3588 | 10.3818 | 26.4863 |
| i | 7.3941 | 11.8541 | 28.6577 |
| j | 7.4265 | 13.1232 | 34.4313 |
| k | 7.1802 | 7.5718 | 19.2626 |
| l | 7.2552 | 10.3865 | 30.7766 |
| m | 7.3167 | 12.3194 | 33.4798 |
| n | 7.3728 | 12.9310 | 34.1983 |
| o | 7.4441 | 13.1217 | 34.4303 |
| p | 7.2523 | 9.4976 | 25.5687 |
| q | 7.3222 | 9.7393 | 25.8062 |
| r | 7.3588 | 10.3818 | 26.4863 |
| s | 7.3941 | 11.8541 | 28.6577 |

[1] is the fusion rule combination form of our method.

As shown in Table 2 and Figure 7, the Curvelet coefficient fusion strategy proposed in this paper, namely the resolution fusion rule used in the Coarse layer and the maximum coefficient fusion rule used in the Detail layer and Fine layer, has the best image fusion effect.

To further demonstrate the effectiveness of the proposed algorithm, the images were fused using different algorithms, including simple average, traditional wavelet fusion and wavelet fusion with resolution constraints. The fusion results were compared with that of the proposed algorithm. The traditional wavelet fusion algorithm applies the mean fusion rule to the low-frequency information of wavelet transform and the maximum coefficient fusion rule to the high-frequency information. In the wavelet fusion with resolution constraints, the resolution fusion rule is applied to the low-frequency information of wavelet transform and the maximum coefficient fusion rule is applied to the high-frequency information.

Table 4 shows the three indices of the four fusion methods, and Figure 8 shows the fusion strip images.

Table 4. Comparison of fusion results of different methods in Area 1.

| Algorithms | Information Entropy | Average Gradient | Spatial Frequency |
|---|---|---|---|
| Our method | 7.6156 | 13.7586 | 35.4629 |
| Wavelet fusion with resolution constraints | 7.3569 | 9.0872 | 28.6397 |
| Traditional wavelet fusion | 7.2260 | 8.2050 | 26.8381 |
| Simple average | 7.1584 | 7.6452 | 19.5543 |

As shown in Table 4, the information entropy, average gradient, and spatial frequency of the proposed algorithm are much greater than those of the other three methods, indicating that the fusion result of the proposed method is the best. By comparing the results of wavelet fusion with resolution constraints and our method, it can be seen that Curvelet fusion achieved better fusion results than wavelet fusion. In addition, based on the value of indices of traditional wavelet fusion and wavelet fusion with resolution constraints, the effectiveness of the resolution fusion rule proposed in this study was demonstrated. It can also be seen intuitively from Figure 8 that the fusion image obtained by our method has better clarity and can show more details.

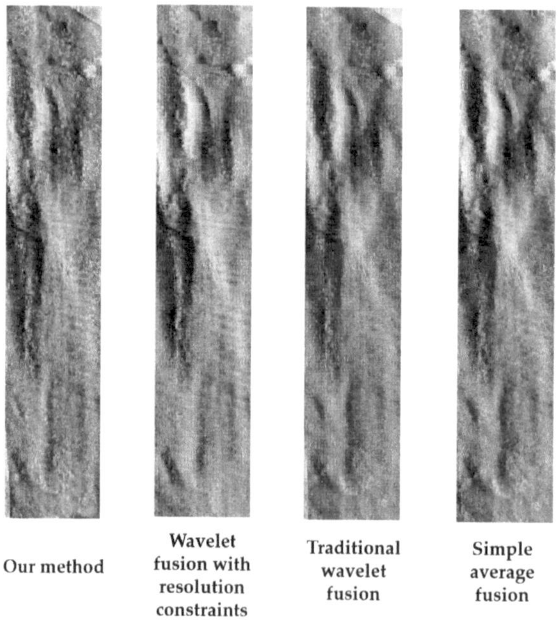


**Figure 8.** Fusion results of the four different methods.

To further verify the effectiveness of the proposed method, the same experiments were repeated for Areas 2 and 3. Figure 9 shows the fusion strip images in Area 2 and Area 3. The evaluation results are shown in Table 5.

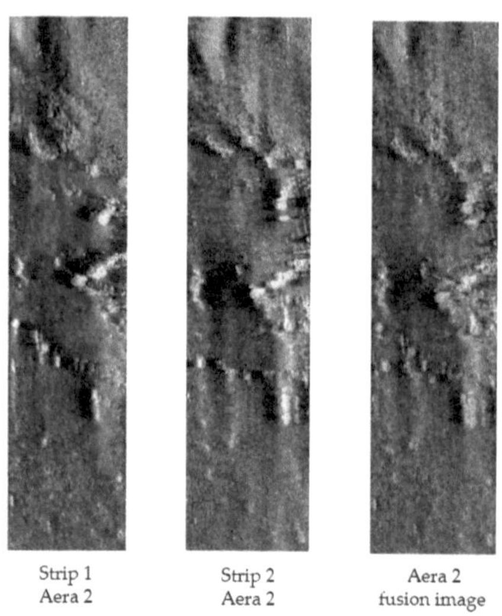


(a)

**Figure 9.** *Cont.*

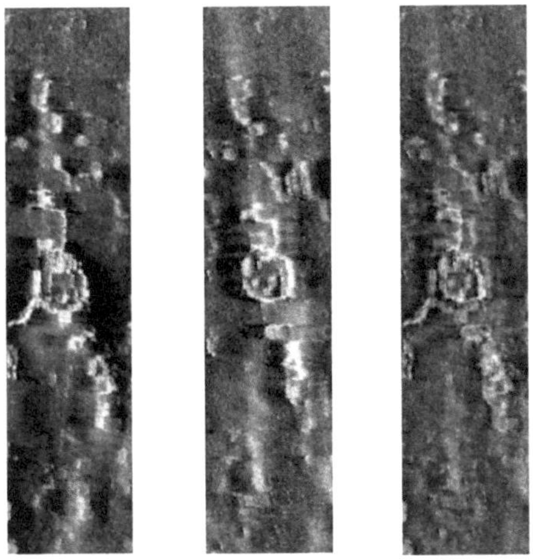


(b)

**Figure 9.** (a) shows the strip images and fusion strip images in Area 2. (b) shows the strip images and fusion strip images in Area 3.

**Table 5.** Comparison of fusion results of different methods in Areas 2 and 3.

| | Ratio | Fusion Algorithms | Information Entropy | Average Gradient | Spatial Frequency |
|---|---|---|---|---|---|
| Area 2 | 0.1428 | Our method | 7.1318 | 11.7527 | 30.4386 |
| | | Wavelet fusion with resolution constraints | 6.9388 | 8.2569 | 25.4472 |
| | | Traditional wavelet fusion | 6.7614 | 7.3451 | 23.7968 |
| | | Simple average | 6.6962 | 6.7457 | 17.0373 |
| Area 3 | 0.2857 | Our method | 7.2367 | 11.8425 | 30.1219 |
| | | Wavelet fusion with resolution constraints | 6.9619 | 7.8150 | 24.4447 |
| | | Traditional wavelet fusion | 6.9174 | 7.4510 | 23.7001 |
| | | Simple average | 6.8657 | 6.8585 | 16.9889 |

As shown in Table 5, the fused images in Areas 2 and 3 of the proposed method have the highest information entropy, average gradient, and spatial frequency, suggesting the best performance in image fusion and validating the effectiveness and stability of the proposed algorithm.

Then, the fused images in the three areas were mosaicked onto the registered strip, which was then rotated clockwise to the original orientation, as shown in Figure 10. Compared with Figure 4c, it can be seen that Figure 9 better reflects the overall characteristics of the features by enhancing detail texture information while retaining the overall trend of the overlapping areas.

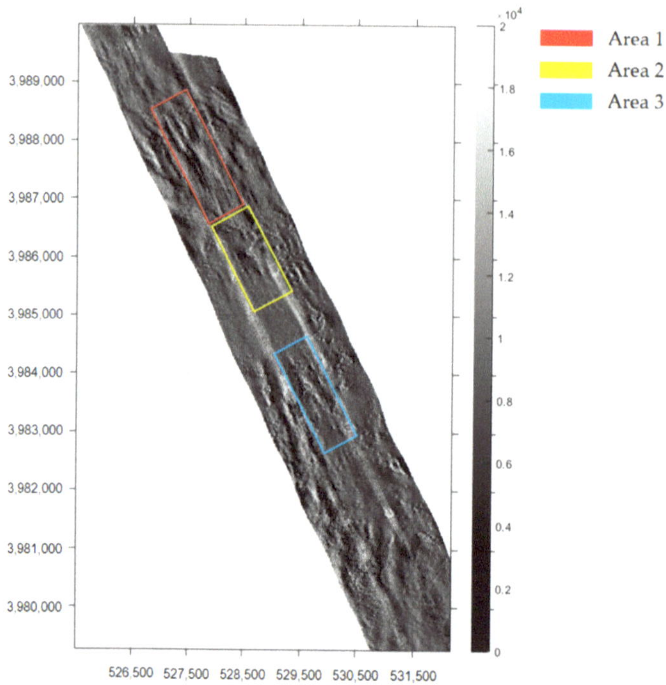


**Figure 10.** Results of strip image mosaicking.

## 5. Conclusions

Current strip image mosaicking algorithms do not consider the influence of the resolution difference of common objects in adjacent images on the results of mosaicking. Moreover, a traditional wavelet fusion algorithm is not able to fully describe the image details. To address these problems, in this study, we proposed an image mosaic method based on Curvelet transform and resolution constraints. Experimental verification using actual measurement data showed that the proposed method can greatly improve the fusion results, which provides high-quality image data for subsequent submarine target recognition and sediment classification, thereby greatly benefiting ocean exploration. However, there are still a lot of improvements to be made in this method, such as human involvement in the process. In view of this, target recognition and other technologies in deep learning can be introduced in the future. Thus, it can automatically identify and extract the areas that need to be fused and achieve full automation.

**Author Contributions:** Conceptualization, S.J. and N.Z.; methodology, S.J. and N.Z.; software, G.B. and Y.C.; validation, L.C., Y.C. and G.B.; formal analysis, S.J.; investigation, N.Z. and L.C.; resources, S.J.; data curation, L.C.; writing—original draft preparation, N.Z.; writing—review and editing, N.Z., S.J. and G.B.; visualization, G.B.; supervision, Y.C.; project administration, S.J.; funding acquisition, S.J. All authors have read and agreed to the published version of the manuscript.

**Funding:** This research was funded by National Natural Science Foundation of China, grant number: 41876103.

**Institutional Review Board Statement:** Not applicable.

**Informed Consent Statement:** Not applicable.

**Data Availability Statement:** The data is kept in_house and not available online.

**Conflicts of Interest:** The authors declare no conflict of interest.

## References


1. Philippe, B. *The Handbook of Sidescan Sonar*; Springer: Berlin/Heidelberg, Germany; New York, NY, USA, 2009; ISBN 978-3-540-42641-7.
2. Fakiris, E.; Papatheodorou, G.; Geraga, M.; Ferentinos, G. An Automatic Target Detection Algorithm for Swath Sonar Backscatter Imagery, Using Image Texture and Independent Component Analysis. *Remote Sens.* **2016**, *8*, 373. [CrossRef]
3. Wang, X.; Zhao, J.; Zhu, B.; Jiang, T.; Qin, T. A Side Scan Sonar Image Target Detection Algorithm Based on a Neutrosophic Set and Diffusion Maps. *Remote Sens.* **2018**, *10*, 295. [CrossRef]
4. Cervenka, P.; Moustier, C.D. Sidescan sonar image processing techniques. *IEEE J. Ocean. Eng.* **1993**, *18*, 108–122. [CrossRef]
5. Chavez, P.S.C., Jr.; Isbrecht, J.A.; Galanis, P.; Gabel, G.L.; Sides, S.C.; Soltesz, D.L.; Ross, S.L.; Velasco, M.G. Processing, mosaicking and management of the monterey bay digital sidescan-sonar images. *Mar. Geol.* **2002**, *181*, 305–315. [CrossRef]
6. Wang, A. *Research on 3D Seafloor Terrian Recovery from the Side Scan Sonar Image*; Wuhan University: Wuhan, China, 2014.
7. Cervenka, P.; Moustier, C.; Lonsdale, P. Geometric corrections on sidescan sonar images based on bathymetry: Application with SeaMARC II and Sea Beam data. *Mar. Geophys. Res.* **1995**, *17*, 217–219. [CrossRef]
8. Cervenka, P.; Moustier, C. Postprocessing and corrections of bathymetry derived from sidescan sonar systems: Application with SeaMARC II. *IEEE J. Ocean. Eng.* **1994**, *19*, 619–629. [CrossRef]
9. Daniel, S.; FL, L.; Roux, C. Side-scan sonar image matching. *IEEE J. Ocean. Eng.* **1998**, *23*, 245–259. [CrossRef]
10. Thisen, E.; Sorensen, H.B.; Stage, B. *Sidescan Sonar Image Matching Using Cross Correlation*; International Society for Optics and Photonics: Bellingham, WA, USA, 2003; Volume 5089.
11. Vandrish, P.; Vardy, A.; Walker, D. Side-scan sonar image registration for AUV navigation. In Proceedings of the 2011 IEEE Symposium on Underwater Technology and Workshop on Scientific Use of Submarine Cables and Related Technologies, Tokyo, Japan, 5–8 April 2011; IEEE: Piscataway, NJ, USA, 2011.
12. Chailloux, C.; Le Caillec, J.M.; Gueriot, D.; Zerr, B. Intensity-Based Block Matching Algorithm for Mosaicing Sonar Images. *IEEE J. Ocean. Eng.* **2011**, *36*, 627–645. [CrossRef]
13. Wang, A.; Zhao, J.; Guo, J.; Wang, X. Elastic Mosaic Method in Block for Side-Scan Sonar Image Based on Speeded-Up Robust Features. *J. Wuhan Univ. Inf. Sci. Ed.* **2018**, *43*, 697–703.
14. Cao, M.; Guo, J. Research on Side Scan Sonar Stripe Image Mosaic Method Based on Common Vision Target. *Geomat. Spat. Inf. Technol.* **2014**, *46–49*, 52.
15. Zhao, J.; Wang, A.; Wang, X. Research on Segment Mosaic Method of Side Scan Sonar Stripe Image. *Geomat. Inf. Sci. Wuhan Univ.* **2013**, *9*, 29–33.
16. Zhao, J.; Shang, X.; Zhang, H. Side-Scan Sonar Image Mosaic Using Couple Feature Points with Constraint of Track Line Positions. *Remote Sens.* **2018**, *10*, 953. [CrossRef]
17. He, F.; Wu, M.; Long, R.; Chen, Z. Accurate Mosaic of side scan sonar image based on SURF feature. *Chin. J. Ocean. Technol.* **2020**, *39*, 35–41.
18. Dharmishtha, P.; Jaliya, U.; Vasava, H. A review: No-reference/blind image quality assessment. *Int. Res. J. Eng. Technol.* **2017**, *4*, 339–343.
19. Yang, X.; Li, F.; Liu, H. A survey of DNN methods for blind image quality assessment. *IEEE Access* **2019**, *7*, 123788–123806. [CrossRef]
20. Wang, Z. Review of no-reference image quality assessment. *Acta Autom. Sin.* **2015**, *41*, 1062–1079.
21. Chern, N.N.K.; Neow, P.A.; Ang, M.H. Practical issues in pixel-based autofocusing for machine vision. In Proceedings of the 2001 ICRA. IEEE International Conference on Robotics and Automation (Cat. No. 01CH37164), Seoul, Korea, 21–26 May 2001; IEEE: Piscataway, NJ, USA, 2001; pp. 2791–2796.
22. Subbarao, M.; Tyan, J. Selecting the optimal focus measure for autofocusing and depth-from-focus. *IEEE Trans. Pattern Anal. Mach. Intell.* **1998**, *20*, 864–870. [CrossRef]
23. Krotkov, E.P. *Active Computer Vision by Cooperative Focus and Stereo*; Springer: New York, NY, USA, 1989; pp. 1–17; ISBN:9781461396659.
24. Chetouani, A.; Beghdadi, A.; Deriche, M. A new reference-free image quality index for blur estimation in the frequency domain. In Proceedings of the 2009 IEEE International Symposium on Signal Processing and Information Technology (ISSPIT), Ajman, United Arab Emirates, 14–17 December 2009; IEEE: Piscataway, NJ, USA, 2009; pp. 155–159.
25. Vu, C.; Phan, T.; Chandler, D. S3: A spectral and spatial measure of local perceived sharpness in natural images. *IEEE Trans. Image Process.* **2012**, *21*, 934–945. [CrossRef]
26. Lu, Y.; Zhang, T.; Zheng, J.; LI, M.; Zhang, C. No-reference blurring image quality assessment based on local standard deviation and saliency map. *J. Jilin Univ. Eng. Technol. Ed.* **2016**, *46*, 1337–1343.
27. Li, Z. *Study on Large-Scale Image Registration Algorithm in Fourier Transform Domain*; Nanjing University of Information Science & Technology: Nanjing, China, 2013.
28. Marichal, X.; Ma, W.; Zhang, H. Blur determination in the compressed domain using DCT information. In Proceedings of the 1999 International Conference on Image Processing (Cat. 99CH36348), Kobe, Japan, 24–28 October 1999; IEEE: Piscataway, NJ, USA, 1999; pp. 386–390.
29. Caviedes, J.; Oberti, F. A new sharpness metric based on local kurtosis, edge and energy information. *Signal Process. Image Commun.* **2004**, *19*, 147–161. [CrossRef]

30. Zhu, Z.; Li, S.; Chen, H. Research on Automatic Focusing Function Based on Image Entropy. *Opt. Precis. Eng.* **2004**, *5*, 537–542.
31. Wang, C.; Cui, L.; Yan, B. Research on the evaluation algorithm of microscopic image sharpness based on Variance-Brenner function. *Equip. Manuf. Technol.* **2020**, *10*, 78–82.
32. Xu, J. *Research on Key Technologies of Side-Scan Sonar Image Mosaic and Segmentation*; East China University of Technology: Fuzhou, China, 2017.
33. Candes, E. Curvelets—A Surprisingly Effective Nonadaptive Representation for Objects with Edges. 2020. Available online: http://citeseerx.ist.psu.edu/viewdoc/summary;jsessionid=0B25E5104FB2CB688F212E84E4F2BD75?doi=10.1.1.161.9294 (accessed on 1 September 2021).
34. Chen, Z. *Curving Transform and Information Extraction Application of Potential Field Data*; China University of Geosciences: Beijing, China, 2012.
35. Vijayalakshmi, G.; Shanthakumar, M.; Muthukumar, K. A Contrast Adjusted Wavelet and Curvelet Transform Based Fusion of Breast Images. *IOP Conf. Ser. Mater. Sci. Eng.* **2020**, *994*, 012027. [CrossRef]
36. Tian, Y.; Zhang, H.; Ma, X. An Image Fusion Method Based on Curvelet Transform and Guided Filter Enhancement. *Math. Probl. Eng.* **2020**, *2020*, 9821715. [CrossRef]
37. Jia, X. *Research on Face Recognition Algorithm Based on Curvwave Transform*; Harbin Engineering University: Harbin, China, 2011.
38. Li, Z.; Li, X.; Ma, L. Research on Definition Assessment based on No-reference Digital Image Quality. *Remote. Sens. Technol. Appl.* **2011**, *26*, 239–246.
39. Xie, X.; Zhou, J.; Wu, Q. A Nonreference Quality Evaluation Index for Image Blur. *Comput. Appl.* **2010**, *4*, 921–924.
40. Gao, F.; Wang, X.; Yang, J.; Zhang, B.; Zhou, H.; Chen, J. Research on Multi-strip Side Scan Sonar Image Fine Mosaic Method. *Sci. Technol. Innov. Appl.* **2021**, *5*, 1–4.

*sensors*

MDPI

*Technical Note*

# Deriving Large-Scale Coastal Bathymetry from Sentinel-2 Images Using an HIGH-Performance Cluster: A Case Study Covering North Africa's Coastal Zone

**Mohamed Wassim Baba [1,2,*], Gregoire Thoumyre [1], Erwin W. J. Bergsma [3] and Christopher J. Daly [4] and Rafael Almar [4]**

[1] CNRS-LEGOS, UMR-5566, 14 Avenue Edouard Belin, 31400 Toulouse, France; gregoire.thoumyre@legos.obs-mip.fr
[2] Center for Remote Sensing Application (CRSA), University Mohammed VI Polytechnic (UM6P), Benguerir 43150, Morocco
[3] French Space Agency (CNES), 31400 Toulouse, France; erwin.bergsma@cnes.fr
[4] IRD-LEGOS, UMR-5566, 14 Avenue Edouard Belin, 31400 Toulouse, France; dalyjchris@gmail.com (C.J.D.); rafael.almar@ird.fr (R.A.)
* Correspondence: wassim.baba@um6p.ma

**Abstract:** Coasts are areas of vitality because they host numerous activities worldwide. Despite their major importance, the knowledge of the main characteristics of the majority of coastal areas (e.g., coastal bathymetry) is still very limited. This is mainly due to the scarcity and lack of accurate measurements or observations, and the sparsity of coastal waters. Moreover, the high cost of performing observations with conventional methods does not allow expansion of the monitoring chain in different coastal areas. In this study, we suggest that the advent of remote sensing data (e.g., Sentinel 2A/B) and high performance computing could open a new perspective to overcome the lack of coastal observations. Indeed, previous research has shown that it is possible to derive large-scale coastal bathymetry from S-2 images. The large S-2 coverage, however, leads to a high computational cost when post-processing the images. Thus, we develop a methodology implemented on a High-Performance cluster (HPC) to derive the bathymetry from S-2 over the globe. In this paper, we describe the conceptualization and implementation of this methodology. Moreover, we will give a general overview of the generated bathymetry map for NA compared with the reference GEBCO global bathymetric product. Finally, we will highlight some hotspots by looking closely to their outputs.

**Keywords:** bathymetry; Sentinel-2; remote sensing; North Africa; HPC

## 1. Introduction

Coastal morphology plays vital role in the global environment. It could be considered as a barrier between land and sea. Coastal regions with shallow water are subject to permanent changes. This impacts their vulnerability to coastal flooding. Moreover, it also can harm economies—e.g., the Netherlands invests on average EUR 35,000,000.00 per year to nourish the coast [1]. Moreover, the depth and shapes of the underwater bathymetry are important for the maritime sectors [2]. Bathymetric maps are also used in biological oceanography, because the depth of the sea is linked to biological characteristics of marine ecosystems. Therefore, generating high-resolution bathymetry maps is an added value to several domains. Despite this major importance, it is very challenging to create bathymetric maps with both accurate spatial resolution and large spatial coverage. It is both time-consuming and very expensive [3]. In recent decades, single-beam echosounding and many other technologies such as multi-beam echosounding, Light Detection and Ranging (LIDAR), have been used to produce the bathymetry of coastal regions at different resolutions and accuracy [3–5]. Classical studies of surveying coastal regions, generally use acoustical techniques which consist of measuring the distance between the device and the bottom of the sea. This technique offers a good accuracy but it is very slow and covers very limited areas. An alternative that can scan a wide area with a good spatial resolution is LIDAR. It uses infrared and green laser transmitter and post-flight data processing techniques to generate survey depths with high accuracy [6,7]. However, this technique is more expensive, especially when the study area is large. To overcome technical and economical issues of in situ observations, the new generation of spaceborne optical remote-sensing sensors could be a good alternative. It is characterized by high spatial resolution and regular revisit time (varying from a few days to 16 days) while covering the entire globe [8]. Several studies are currently moving in this direction, and try to use remote sensing data to retrieve the water depth in coastal regions. Hamylton et al. [9] compared two approaches to estimate the bathymetry at 5 m of resolution. This study was established at both the Lizard Island and Sykes Reef sites, by using WorldView-2 images. Chybicki [10] demonstrated that an inversion of Sentinel 2A/B radiance is useful to derive an estimation of the shore bathymetry (for areas with a depth lesser than 18 m in South Baltic coastline. Other researchers used similar method to derive the bathymetry such as [11,12]. This method is efficient for nearshore areas but has a principal limitation of this method is the water turbidity which degrades the quality of the results. Thus, the objective of this article is to present a large-scale implementing an High Performance Cluster (HPC) methodology to derive the bathymetry from high spatial resolution images (of type Sentinel 2A/B) from a regional, continental to global scale. We present an approach that processes many images simultaneously with a computation time around 1.5 h per image per CPU (central processing unit). At first stage, we describe physical laws that control the estimation of the water depth followed by a detailed HPC-workflow to implement the bathymetry derivation-code. In the results section we present a study case applied in North African's coasts. Detailed description of computational time and memory while running the code over the tiles will be given. Finally, in discussion section we discuss the limitations of the methodology, and present a recommendation to run it over the world with specific IT resources.

## 2. Study Area and Data Source

### 2.1. Study Area

The arid coasts of North Africa, extending over more 6000 km from Morocco Atlantic Coast to the Nile Delta (Figure 1), are undergoing pronounced shoreline retreats and coastal flooding that are reported as a consequence of the ongoing sea level rise resulting from global warming. The coastal zone plays an important role in the economic development of this region [13]. Indeed, more than 60% of the population live in coastal cities, and 90% of the country's industries are based along the coast [14,15]. Of particular interest are the abnormal shoreline dynamics for deltaic and sandy beaches, which are severely impacted by abrupt decadal variabilities in both climatic and anthropogenic drivers resulting in their increased vulnerability to disturbances from coastal hazards. Unfortunately, the evolution, distribution and impacts of these drivers remain largely unquantified, let alone understood, for these extensive arid coasts that harbor the major portion of North Africa's population as well as unique and fragile marine ecosystems.

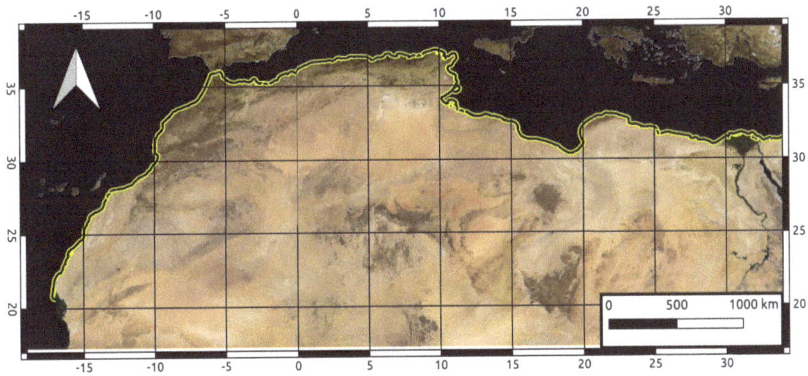


**Figure 1.** Contextual map of our study area: coasts of North Africa are represented by the yellow buffer. Reference coordinate system used: World Geodetic System (WGS84).

### *2.2. Sentinel 2A/B Images Retrieval*

Sentinel 2A and 2B are two twin polar-orbiting satellites launched in 2015 and 2017, respectively, as part of the European Copernicus program [16]. They are designed for the operational monitoring of atmosphere, land and ocean. With a wide-swath and a multispectral imager (MSI) with 13 spectral bands (from 443 nm to 2190 nm), a high spatial resolution imagery varying from 10 m for the majority of the bands covering the visible, and very near infrared (VNIR), to 60 m for the short wave infrared (SWIR). In addition to this variety of bands and the high spatial resolution, Sentinel 2A/B has a temporal resolution reaching a maximum of 5 days (more you are far than equator less the temporal resolution [17]). They allow the observation and monitoring of land, atmosphere and ocean every 5 days.

In this study, 287 Sentinel 2A/B tiles cover the North Africa coastal zone (Figure 1) were used. First, we queried all available Sentinel 2A/B (S2-L1C) images between January 2015 and January 2020 in the PEPS collection of the Theia Land data center. For the analysis in this work we limit the computations to ten scenes of cloud-free Sentinel 2A/B images during a 5 year span from 2015 to 2020. A sensitivity analysis in the original method article [18] and for West-African hotspots [19], shows that the wave power (as a function of wave height and period) as well as directional spread of the wave field affect depth estimation. Generally, the more powerful, narrow-spread wave fields allow for a more accurate estimation of the water depth.

Considering this, 10 cloud free images with most powerful waves are selected in this study. Since we are working on 287 tiles and in each tile there are 10 images, we integrated 2870 images in the HPC. If our area (the selected image) is totally covered by water, it has an execution time of 72 core-hours (36 CPUs used for two hours). If the image has a terrestrial part this time decreases. This is why we fixed walltime for running the Portable Batch System (PBS) in 2 h.

#### Global Water Mask

Since it is not our objective to estimate water depths on lands, the land is filtered from the image by using a land-mask. The set of routines is fed by a water mask which not only prohibits depth estimation at land but it is also used to compute the distance to the shore that is used in adaptive tile-sizes during the computation. Here we use an existing global water mask with an initial resolution of 300 m. It was generated by European Space Agency Climate Change by combining SAR images and recent Landsat-derived products. The water mask was re-sampled to a resolution of 500 m in order to be compatible with our outputs: water depths are estimated on 500 m resolution.

## 3. Methods

### 3.1. Physical Description

Ocean waves, when they are about to arrive in the coastal zone, are free moving waves. Only in intermediate $h_{int} = \frac{L}{2}$ to shallow water $h_{sh} = \frac{L}{20}$, the propagation of these waves is limited by, among other physical processes, the water depth. As the water depth reduces towards shore, waves increasingly "feel" the bottom by increasing bottom-friction until the waves eventually break close to shore. Depth domination of the wave propagation can be described with through a mathematical relationship; the dispersion relation for free surface waves.

$$c^2 = \frac{g}{k}\tanh(kh) \Leftrightarrow h = \frac{\tanh^{-1}\left(\frac{c^2 k}{g}\right)}{k} \quad (1)$$

To solve (1), one is after a pair of wave celerity ($c$), wave length ($L$) or number ($k$) and wave period ($T$) or frequency ($\omega$). Here we use the approach following Bergsma et al. [18] to find wave phase shifts (leading to celerity $c$) per wave number ($k$) in different detector-bands using a Radon-Transform based Fourier slicing techniques. The great benefit of this Radon-Transform based technique is the limited dependence on image resolution to estimate wave propagation while maintaining computation performance.

### 3.2. Numerical Implementation and Regional Application

#### 3.2.1. IT Workflow

To minimize the execution time, each Sentinel 2A/B image is split into 36 subsets (6 × 6). This ensure that all the available CPUs are used at the same time. Thereafter, we move a sub-window over the pixels that we want to process (depending on the output resolution). In this sub-window the physical equations described in Section 3.1. The approach of this decomposition is illustrated in Figure 2. The size of each sub-window depends primarily on the distance to the shore. It's equation is given as follows:

$$\kappa = \min\left(2.5, \left[\frac{\log_{10}(D_{shore})}{2}\right]^2\right) \quad (2)$$

$$W_{xs} = \kappa \times L_{win} \quad (3)$$

where, $\kappa$ is the factor regulating the size of the window. $D_{shore}$ is the distance to the shore and $W_{xs}$ represents the window length (The size of the window is $W_{xs} \times W_{xs}$. This widow is applied for each study bands. $L_{win}$ is constant fixed in 200 m in this study.

Once the pixel is surrounded by the window, a Radon Transform (RT) is applied as described in [18]:

$$R_{sub_I} = \text{radon}(sub_I) \quad (4)$$

$R_{sub_I}$ represents the sinogram calculated over the sub-window domain. This sinogram allows to extract the main direction of the wave by taking the maximum variance or standard deviation over all directions. Once the direction is found, a fast-Fourier transform DFT is performed over beam with the selected angle to obtain the phase per band ($\phi$) (following the methodology of [18]). Using two images with a slight $\delta(t)$, one can compute the phase shift $\Phi$ and find celerity ($c$) by linking $\lambda$, $\Phi$ using (5)

$$c = \frac{\Delta\Phi}{2\pi\lambda\delta(t)} \quad (5)$$

where, $\delta(t)$ represents the difference time in seconds (s) between the acquisition detector bands. Here only the 10-m resolution bands are used, but possible Radon-augmentation enables the use of all bands [18].

The extract the depth ($D$) a numerical implementation of Equation (1) is implemented. A graphical description of the algorithm that summarizes the workflow is given in Figure 3. The IT implementation of the physical code is presented in Section 3.1.

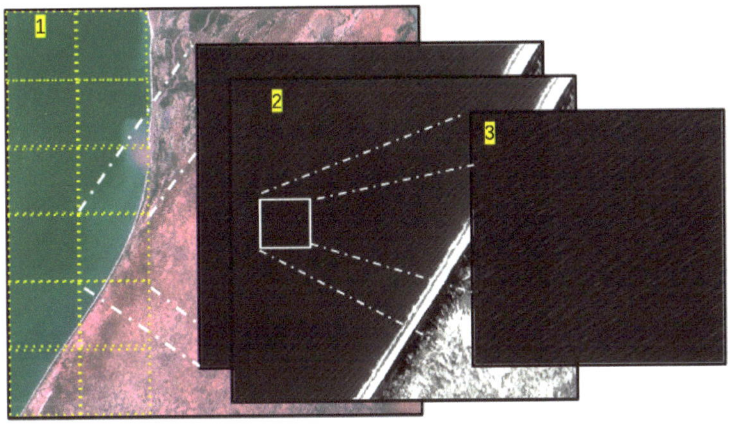


**Figure 2.** (1). Description of the manner by which the slicing is performed depending on the number of CPUs. (2). Creating a sub-window over the point of interest. (3). The sub-window where the different variables (e.g., celerity and depth) are computed.

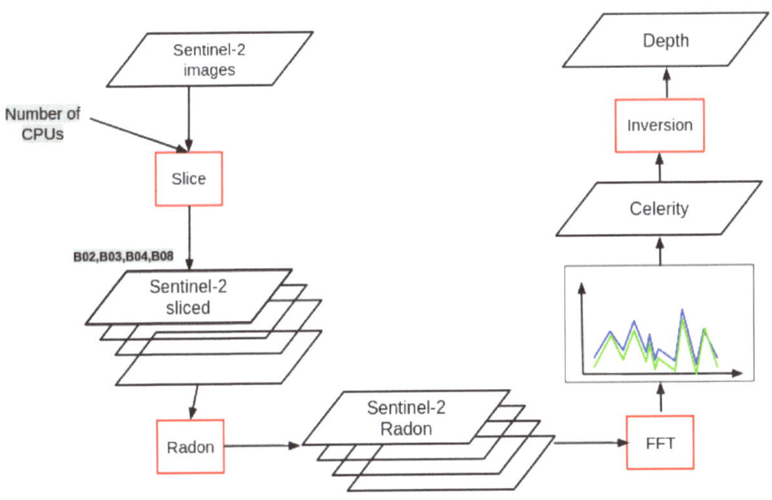


**Figure 3.** Workflow for Sentinel 2A/B process, from the acquisition to retrieving the bathymetry.

3.2.2. HPC Implementation

A High Performance Computing (HPC) techniques refers to a supercomputer with a high level of performance as compared to a general-purpose computer. The basic elements of HPC techniques consists of node, queue, and a job. A node is a single physical or logical computer with one or more processors. Queuing systems are responsible for managing job requests which are shell generally scripts submitted by users. A job is a collection of instructions that a user initiates. Each job reserves specific resources in term of Random-access memory (RAM) and CPUs. To sum up, the computations are performed by the cluster, by submitting a job request to a specific batch queue. The scheduler will assign your job to a compute node in the order determined by the policy on that queue and the availability of an idle compute node.

In our case, the different simulations were established in the CNES-HPC cluster. All available cores are allocated with 36 Central Processing Units (CPU) and 70 Gb of RAM. The data are collected from the datalake; thereafter, each node treat separately one image (Figure 4).

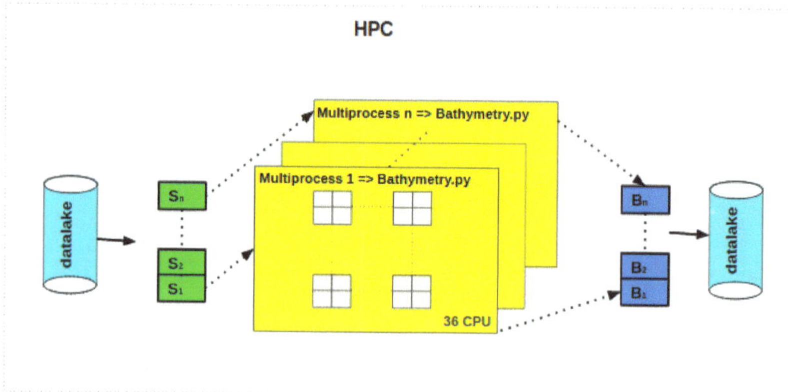


**Figure 4.** Schematic overview of a HPC architecture. From the right to the left: loading of the data from the datalake Sentinel-1, 2, ..., n. Then in each node (the yellow box) the image is split into 36 sub-images that are treated in parallel. Each output is merged and then saved in the datalake.

### 3.3. *North Africa Coastal Bathymetry Showcase*

Water depth maps are at 500 m of spatial resolution. Figure 5 shows the result of the computation for the North Africa coastal region. Bathymetry is computed until the 2% of potential estimated depth [17] using ERA5 hindcast (ECMWF). This covers most of the shallow shelves, bays (Gabes bay) and delta (Nile). This potential varies from shallow in the Mediterranean Sea, due to limited fetch to deeper waters along the open Atlantic Coast. S2Shores results show in general a good agreement with GEBCO, the referent global bathymetry product, combining numerous surveys and previous altimetry-based coarse satellite estimations.

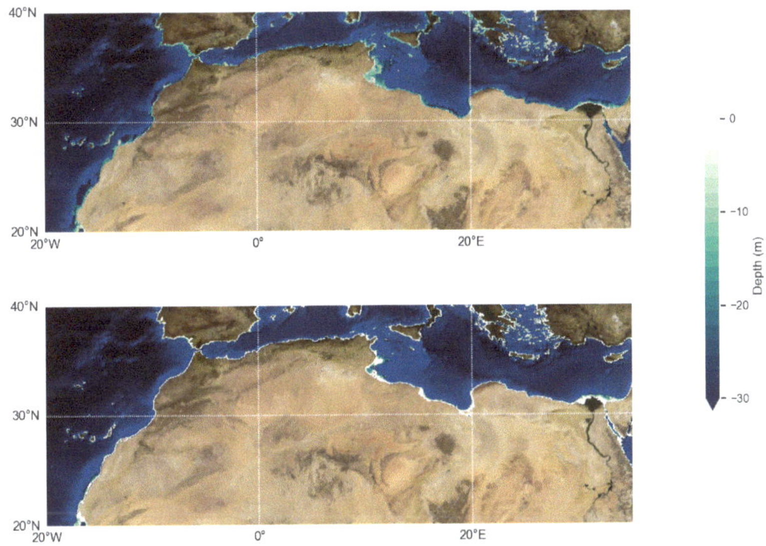


**Figure 5.** Bathymetry along the North Africa coastline with our S2 shores estimate at the top and the reference GEBCO global product below.

However, it was found that GEBCO has substantial issues in coastal and shallow waters, and often generates some bumps due to interpolation of different source data-set and ends up at the shore with a linear interpolation to the shoreline. A more thorough validation from an unrivaled local survey remains invaluable in terms of ground truth and is yet to be done in this area. However, [19] conducted in Senegal a comparison with echo-sounder surveys and our estimated showed an accuracy of meters (RMSE is varying between 2 and 5 m, i.e., 10–20%).

The implemented method allow to represent different types coasts varying from shallow depth to deep depth as shown in Figure 6 for the three different coasts in Morocco; Djerba island, Tunisia; and Egypt. In these cases the maximum depth, or deep water limit is reached around 45 m.

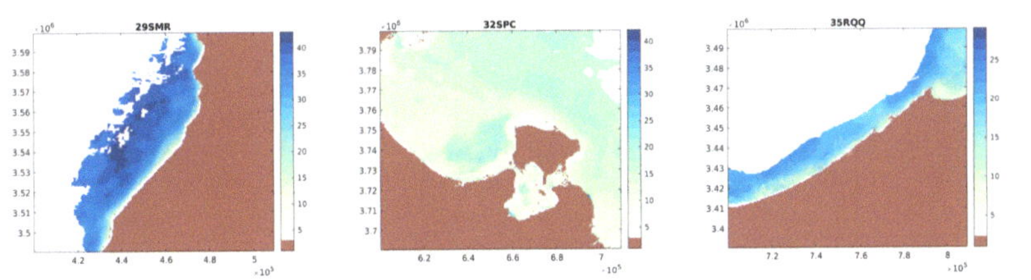


**Figure 6.** Illustration of S2Shores satellite-derived coastal bathymetry at showcases zones. The the tiling grid IDs for Sentinel 2 are given at the top of each image.

## 4. Conclusions and Way Forward

By positioning itself on a global coverage, S2SHORES covers different coastal zone environments around the world with a sensor resolution of 10 m (Sentinel 2A/B) over a variable band from the shore and down to a depth limit of about 50 m [19]. The approach can be extended to other missions (Pleiades, World-View, SPOT6, etc.) of higher resolution for high precision need at hot-spots [20]. Sentinel 2A/B assets are the free data and the long-term nature of the Sentinel 2A/B program. Indeed, Sentinel-2A and B cover the periods 2015–2022 and 2017–2024, respectively, and their successors Sentinel-2 C and D to ensure the continuity of the mission are already scheduled for the next seven years. We show here that the recent availability of new global high resolution products such as Sentinel 2A/B (ESA) COPERNICUS constellation, combined with the latest methodology of bathymetry retrieval, can be applied globally: offering a new vision with uniform method. The results presented here are a pathway toward new EOS coastal products. Besides the approach employed and the scores, the new possibilities are evident. There is a current strong and increasing demand for such global product, after the decades old supremacy of state-of-the-art global relatively coarse resolution and rather inappropriate at the coast. This for coastal engineering, management and planning, risk forecasting and mitigation abut also scientific advance. The way forward include the optimal use of the regular, and increasing, revisit of these earth observation satellite to monitor coastal changes under climate change.

**Author Contributions:** M.W.B. and G.T. did the analyses and wrote the manuscript; R.A., E.W.J.B. and C.J.D. participated to the ideas, revised the manuscript at different stages and collaborated on the ideas. All authors have read and agreed to the published version of the manuscript.

**Funding:** Several sources were used to achieve this work. We thank especially CNES and IRD.

**Data Availability Statement:** The S2Shores maps are available at CNES datalake. Please contact R.A. (rafael.almar@ird.fr) if you are looking for their use.

**Acknowledgments:** We thank PEPS team at CNES for delivering Sentinel 2A/B images on the datalake.

**Conflicts of Interest:** The authors declare no conflict of interest.

## References

1. Jonkman, S.N.; Hillen, M.M.; Nicholls, R.J.; Kanning, W.; van Ledden, M. Costs of Adapting Coastal Defences to Sea-Level Rise—New Estimates and Their Implications. *J. Coast. Res.* **2013**, *29*, 1212–1226. [CrossRef]
2. Guenther, G.C. Airborne lidar bathymetry. *Digit. Elev. Model Technol. Appl. DEM Users Man.* **2007**, *2*, 253–320.
3. Irish, J.L.; White, T.E. Coastal engineering applications of high-resolution lidar bathymetry. *Coast. Eng.* **1998**, *35*, 47–71. [CrossRef]
4. Heezen, B.C.; Tharp, M. World Ocean Floor Map. Available online: https://www.lib.uchicago.edu/collex/exhibits/marie-tharp-pioneering-oceanographer/1977-world-ocean-floor-map/ (accessed on 16 October 2021).
5. Barton, C. Marie Tharp, oceanographic cartographer, and her contributions to the revolution in the Earth sciences. *Geol. Soc. Lond. Spec. Publ.* **2002**, *192*, 215–228. [CrossRef]
6. Guenther, G.C.; Thomas, R.W.; LaRocque, P.E. Design considerations for achieving high accuracy with the SHOALS bathymetric lidar system. In *CIS Selected Papers: Laser Remote Sensing of Natural Waters: From Theory to Practice*; International Society for Optics and Photonics: Bellingham, WA, USA, 1996; Volume 2964, pp. 54–71.
7. Irish, J.L.; McClung, J.; Lillycrop, W.J. Airborne Lidar Bathymetry: The SHOALS System. Available online: https://trid.trb.org/view/652843 (accessed on 16 October 2021).
8. Cazenave, A.; Le Cozannet, G.; Benveniste, J.; Woodworth, P.; Champollion, N. Monitoring Coastal Zone Changes from Space. Available online: https://eos.org/opinions/monitoring-coastal-zone-changes-from-space (accessed on 16 October 2021).
9. Hamylton, S.M.; Hedley, J.D.; Beaman, R.J. Derivation of high-resolution bathymetry from multispectral satellite imagery: A comparison of empirical and optimisation methods through geographical error analysis. *Remote Sens.* **2015**, *7*, 16257–16273. [CrossRef]
10. Chybicki, A. Mapping south baltic near-shore bathymetry using Sentinel-2 observations. *Pol. Marit. Res.* **2017**, *24*, 15–25. [CrossRef]
11. Stumpf, R.P.; Holderied, K.; Sinclair, M. Determination of water depth with high-resolution satellite imagery over variable bottom types. *Limnol. Oceanogr.* **2003**, *48*, 547–556. [CrossRef]
12. Lyons, M.; Phinn, S.; Roelfsema, C. Integrating Quickbird multi-spectral satellite and field data: Mapping bathymetry, seagrass cover, seagrass species and change in Moreton Bay, Australia in 2004 and 2007. *Remote Sens.* **2011**, *3*, 42–64. [CrossRef]
13. Hzami, A.; Heggy, E.; Amrouni, O.; Mahé, G.; Maanan, M.; Abdeljaouad, S. Alarming coastal vulnerability of the deltaic and sandy beaches of North Africa. *Sci. Rep.* **2021**, *11*, 2320. [CrossRef]
14. Snoussi, M.; Ouchani, T.; Niazi, S. Vulnerability assessment of the impact of sea-level rise and flooding on the Moroccan coast: The case of the Mediterranean eastern zone. *Estuar. Coast. Shelf Sci.* **2008**, *77*, 206–213. [CrossRef]
15. Hakkou, M.; Maanan, M.; Belrhaba, T.; El Khalidi, K.; El Ouai, D.; Benmohammadi, A. Multi-decadal assessment of shoreline changes using geospatial tools and automatic computation in Kenitra coast, Morocco. *Ocean Coast. Manag.* **2018**, *163*, 232–239. [CrossRef]

16. ESA Sentinel-2 Delivers First Images. Available online: https://www.esa.int/Applications/Observing_the_Earth/Copernicus/Sentinel-2/Sentinel-2_delivers_first_images (accessed on 16 October 2021).
17. Bergsma, E.W.J.; Almar, R. Coastal coverage of ESA' Sentinel 2 mission. *Adv. Space Res.* **2020**, *65*, 2636–2644. [CrossRef]
18. Bergsma, E.W.J.; Almar, R.; Maisongrande, P. Radon-Augmented Sentinel-2 Satellite Imagery to Derive Wave-Patterns and Regional Bathymetry. *Remote Sens.* **2019**, *11*, 1918. [CrossRef]
19. Daly, C.J.; Baba, W.; Bergsma, E.; Almar, R.; Garlan, T. The New Era of Regional Coastal Bathymetry from Space: A Showcase for West Africa using Sentinel-2 Imagery. Available online: https://eartharxiv.org/repository/view/187/ (accessed on 16 October 2021).
20. Almar, R.; Bergsma, E.W.; Maisongrande, P.; de Almeida, L.P.M. Wave-derived coastal bathymetry from satellite video imagery: A showcase with pleiades persistent mode. *Remote Sens. Environ.* **2019**, *231*, 111263. [CrossRef]

MDPI AG
Grosspeteranlage 5
4052 Basel
Switzerland
Tel.: +41 61 683 77 34

*Sensors* Editorial Office
E-mail: sensors@mdpi.com
www.mdpi.com/journal/sensors

www.ingramcontent.com/pod-product-compliance
Lightning Source LLC
Chambersburg PA
CBHW041346210926
51850CB00013B/3002
* 9 7 8 3 7 2 5 8 4 3 5 5 8 *